Clinical Trial Registries:
A Practical Guide for
Sponsors and Researchers
of Medicinal Products

Edited by MaryAnn Foote

Birkhäuser Verlag
Basel · Boston · Berlin

MaryAnn Foote
Abraxis BioScience, Inc.,
11777 San Vincente Blvd
Suite 550
Los Angeles, CA 90049
USA

Library of Congress Cataloging-in-Publication Data

Clinical trial registries : a practical guide for sponsors and researchers of medicinal
products / edited by MaryAnn Foote.
  p. cm.
  Includes bibliographical references.
  ISBN 978-3-7643-7578-2
  ISBN 3-7643-7578-7 (alk. paper)
  1. Clinical trials--Reporting. I. Foote, MaryAnn.

  R853.C55C576 2006
  615-5072'4--dc22

                    2006047741

Bibliographic information published by Die Deutsche Bibliothek
Die Deutsche Bibliothek lists this publication in the Deutsche Nationalbibliografie;
detailed bibliographic data is available in the Internet at <http://dnb.ddb.de>.

ISBN   3-7643-7578-7 Birkhäuser Verlag, Basel – Boston – Berlin
ISBN   978-3-7643-7578-2

The publisher and editor can give no guarantee for the information on drug dosage and
administration contained in this publication. The respective user must check its accuracy
by consulting other sources of reference in each individual case.
The use of registered names, trademarks etc. in this publication, even if not identified as
such, does not imply that they are exempt from the relevant protective laws and regulations or free for general use.

© 2006 Birkhäuser Verlag, P.O. Box 133, CH-4010 Basel, Switzerland
Part of Springer Science+Business Media
Printed on acid-free paper produced from chlorine-free pulp. TCF ∞

ISBN   3-7643-7578-7   ISBN 3-7643-7583-3 (eBook)
ISBN   978-3-7643-7578-2

9 8 7 6 5 4 3 2 1   www.birkhauser.ch

Contents

List of contributors

Tracy Beck, Eli Lilly and Company, Lilly Corporate Center, Indianapolis, IN 46285, USA; e-mail: becktj@lilly.com

Lucy Erdelac, PIA-Astrolabe Analytica, A Division of Thomson Scientific, Inc., 101 Gibraltar Rd, Suite 200, Horsham, PA 19044, USA; e-mail: lucy@erdelac.com

MaryAnn Foote, Abraxis BioScience, Inc., 11777 San Vincente Blvd, Suite 550, Los Angeles, CA 90049, USA; e-mail: mfoote@abraxisbio.com

Kenneth Getz, Tufts Center for the Study of Drug Development, Tufts University, 192 South Street, Suite 550, Boston, MA 02111, USA; e-mail: kenneth.getz@tufts.edu

Alan Goldhammer, PhRMA, 950 F Street, NW, Washington, DC 20004, USA; e-mail: agoldhammer@phrma.org

Charlotte Haug, Norwegian Medical Journal (Den norske lægeforening), Akersgata 2, 0107 Oslo, Norway; e-mail: charlotte.haug@legeforeningen.no

Mark Jungemann, Pharmaceutical Project Management – Product Phase Eli Lilly and Company, Lilly Corporate Center, Indianapolis, IN 46285, USA; e-mail: mej@lilly.com

Karmela Krleža-Jerić, Randomised Controlled Trials Unit, Canadian Institutes of Health Research, 160 Elgin Street, 9th floor, Address Locator 4809A, Ottawa, Ontario, K1A 0W9, Canada; e-mail: kkrleza-jeric@cihr-irsc.gc.ca

Takahiro Kiuchi, University Hospital Medical Information Network, The University of Tokyo, Tokyo, Japan; e-mail: tak-kiuchi@umin.ac.jp

Lawrence E. Liberti, PIA-Astrolabe Analytica, A Division of Thomson Scientific, Inc., 101 Gibraltar Rd, Suite 200, Horsham, PA 19044, USA; e-mail: lawrence.liberti@thomson.com

Ana Marušić, Croatian Medical Journal, Zagreb University School of Medicine, Šalata 3, 10000 Zagreb, Croatia; e-mail: marusica@mef.hr

Hisako Matsuba, EPS Co., Ltd., 2-23 Shimomiyabi-cho, Shinjuku-ku, Tokyo, 162-0822, Japan; e-mail: h.matsuba@eps.co.jp

Dan McDonald, Thomson CenterWatch, 22 Thomson Place, 47F1, Boston, MA 02210, USA; e-mail: daniel.mcdonald@thomson.com

Yasuo Ohashi, School of Health Sciences and Nursing, The University of Tokyo, Tokyo, Japan; e-mail: ohashi@epistat.m.u-tokyo.ac.jp

Jean Papaj, Thomson Scientific, Inc., Philadelphia, Pennsylvania, USA; e-mail: jean.papaj@thomson.com

Tim Peoples, Cyberonics, Inc., 100 Cyberonics Blvd, Houston, TX 77058, USA; e-mail: timpeoples@gmail.com

Susan Siefert, Cyberonics, Inc., 100 Cyberonics Blvd, Houston, TX 77058, USA; e-mail: ses@cyberonics.com

Kiichiro Tsutani, Department of Pharmacoeconomics, Graduate School of Pharmaceutical Sciences, The University of Tokyo, Tokyo, Japan; e-mail: tsutani-tky@umin.ac.jp

Eiji Uchida, Second Department of Pharmacology, School of Medicine, Showa University, Tokyo, Japan; e-mail: uchieiji@med.showa-u.ac.jp

Steve Zisson, Thomson CenterWatch, 22 Thomson Place, 47F1, Boston, MA 02210, USA; e-mail: stephen.zisson@thomson.com

Preface

Science, and the scientific method in particular, comprises several steps: observation, questioning, formulation of a hypothesis, testing and retesting the hypothesis, and dissemination of the results. All steps must be completed. This book, *Clinical Trial Registries: A Practical Guide for Sponsors and Researchers of Medical Products*, is about science, in particular the public statement of the hypothesis of a clinical trial and the public dissemination/publication of the results of the trial.

Much has been written about the alleged lack of transparency in clinical trials and the subsequent lack of publication of the results, and many of the chapters have detailed information and references on this topic.

Many readers will be familiar with Food and Drug Administration Modernization Act section 113 (FDAMA 113) and many may believe that FDAMA 113 is the primary authority on the topic. FDAMA 113, which is discussed in several chapters, was initiated in 1997 and established a registry in the United States of America for clinical trials that evaluated the efficacy of treatments for severe or life-threatening diseases. Originally, the database was established to address the medical issue of cancer and of HIV/AIDS and was expanded to other areas of disease. FDAMA 113 called for registration of a trial within 21 days of starting enrollment and required regular updating of the information in the registry. A Web site, clinicaltrials.gov, a service of the National Institutes of Health (NIH), was started in 1998 in response to FDAMA 113. Originally, clinicaltrials. gov only accepted investigational new drug (IND) studies for severe and life-threatening diseases, but the criteria were

broadened in 2004 to encompass other studies and to fulfill
the International Committee of Medical Journal Editors (IC-
MJE) new requirement of trial registration at study inception
for subsequent consideration for publication. Several organizations, including the Pharmaceutical Research and Manufacturers of America (PhRMA) and the World Health Organization
(WHO), have become involved in trying to establish guidelines
and Web sites for both the registration of clinical trials and the
dissemination of results. However, complete agreement has
not been reached among regulators, clinical trial investigators,
editors, and the pharmaceutical industry, who do not agree on
what should be revealed in a clinical trial registry, when information should be revealed, and how results should be disseminated beyond a peer-reviewed journal article. Add in legislation or pending legislation by at least two states in the United
States (Maine and Virginia) and legislation pending in other
states and countries, including Canada, Australia, Japan, and
the European Union, the need for a single resource, i.e., this
book, to frame the issues is evident.

The book contains much valuable information and many
points of view. Consider the so-called '20 fields for a clinical trial registry'. One chapter written by ICMJE members
(Marušić and Haug) discusses the journal editors' perspective
on clinical trial registration and the need for 20 proposed fields
to prevent publication of only part of the data collected in a
trial. The point of view of the Ottawa Group (Krleža-Jerić) is
that the data enclosed in the 20 fields of a clinical trial registry
does not go far enough and that the pharmaceutical industry
needs to be more open about its clinical trials. The pharmaceutical industry perspective is provided in another chapter
(Beck) that discusses clinical trial registries and the publication of results, and the potential problems that drug sponsors
face with automatic completion of the 20 fields. The author

suggests that some in the pharmaceutical industry support the idea that information can be prospectively added to a clinical trial registry, but remain 'locked' from the general public (and competitors). At the time of submission of a manuscript based on the clinical trial, journal editors would have access to the information and could thus verify that the endpoints reported were indeed the endpoints prospectively planned. The pharmaceutical industry must protect proprietary information, lest innovation be stifled, and has committed to registering all information at trial inception unless doing so jeopardizes the drug sponsor (Goldhammer). In this case, the information would be revealed at a later date.

Efforts have been made by the pharmaceutical industry to comply with rules, regulations, and guidelines (McDonald and Zisson). Compliance has not been without a cost: some companies report 25 to 40 full-time staff were needed to ensure clinical trial transparency.

PhRMA has been instrumental in establishing a publicly accessible Web site, which can be used for the dissemination of results from clinical trials, www.clinicalstudyresults.org (Goldhammer). PhRMA members have committed to registration of clinical trials and dissemination of results, positive and negative.

Lest we lose track of patients and their needs, one of the reasons given for the need for clinical trial registries and the primary reason for the establishment of clinicaltrials.gov, a chapter focuses on the topic of what patients want and need in a registry and provides results of original research (Getz). Patients appear to have very distinct needs and wishes that may not be addressed by either ICMJE or industry perspectives.

Another chapter (Liberti, Erdelac, and Papaj) suggests that the time may be right for a new type of clinical trial registry, an upgraded edition 2.0 version, as it were. An ongoing problem with clinical trial registries is the plethora of sites. At this time, no one site appears to satisfy the needs of the various stakeholders. A quick perusal of Web sites suggests that no uniform format exists that allows patients and their healthcare providers or journal editors for that matter, to easily compare and understand the information in the registries.

The issue of clinical trial transparency is an international one, and different countries have addressed the issue in different ways. In the United States, clinicaltrials.gov has been in existence for approximately 8 years and is the registry chosen by the ICMJE as the preferred registry. Other registries do exist to meet the needs of patients in other parts of the world. One chapter in the book offers thoughtful discussion of the state of clinical trials, clinical trial registries, and publication of results in Japan (Matsuba, Kiuchi, Tstutani, Uchida, and Ohashi).

Expert guidance in how to approach establishing an in-house registry is given (Jungemann). A project management approach was taken by Eli Lilly, who maintains a Web site considered to be among the best Web sites available for information about its trials and products.

To increase the usefulness of the book, two appendices are included: one provides Web addresses for all known international governmental, oncology cooperative, industry, and commercial databases for clinical trial registration or the posting of results, or both (Foote). Also included is an annotated bibliography of important references concerning the topic of clinical trial registries (Peoples and Siefert).

The story of clinical trial registries and dissemination of clinical trial result is not finished, and this book does not supply all the answers. Much work remains, thoughtful work that ultimately produces a way for patients to learn about trials, provides biopharma companies a way to comply with ICMJE and trade group guidelines without jeopardizing intellectual property, and allows journal editors a way to be confident that bias is not introduced into the published results of clinical trials. The authors of the chapters have done their best to help the story move forward.

The reader will note some overlap and repetition of information among the chapters. The overlap has been maintained to provide different points of view and to foster further discussion. Care has been taken to ensure the accuracy of the data in this book, including attribution to original authors. Neither the publisher nor the editor can be responsible for any errors contained within. All opinions stated are those of the individual authors.

No scientist works in a vacuum and I acknowledge years of fruitful discussions with Dr. Laurence Hirsch and Dr. Philip David Noguchi. I am grateful to Dr. Larry Bell for allowing me time and resources to delve deeply into the world of clinical trial registries and all it entails. I thank Dr. Hans-Detlef Klüber of Birkhauser for realizing the need for this book and working quickly with his staff to make it happen.

MaryAnn Foote, PhD
Westlake Village, California, April 2006

Clinical trial registries and publication of results – a primer

MaryAnn Foote

Abraxis BioScience, Inc., 11777 San Vincente Blvd, Suite 550,
Los Angeles, CA 90049, USA

Introduction

The issue of clarity in clinical trials is not new and many of the chapters in this book elegantly detail various aspects of the issue. This chapter summarizes some of the published information on the topic of clinical trial registries and publication of results. The reference section of each chapter can be used to find more information on specific topic, and Appendix 2 is a compilation of a wide variety of published information on the topic.

One of the first papers stating concerns about potential publication bias was published in 1986 by Simes [1], who suggested bias in reporting positive or promising results. Simes reviewed the published literature of the time concerning survival in patients with advanced ovarian cancer or multiple myeloma. In the setting of advanced ovarian cancer, the available published literature suggested a survival advantage for combination chemotherapy, while no significant difference in survival was noted in an analysis of large cooperative trials. In the setting of multiple myeloma, a pooled analysis of published trials also suggested a survival advantage for combination chemotherapy, which was confirmed by analysis of cooperative group trials, but with a reduced estimated magnitude of the benefit. Chalmers [2] extended the concerns of Simes by stating that underreporting of clinical research could be considered to be scientific fraud.

Heightened renewed interest in clinical trial reporting started approximately 2001, with the pharmaceutical industry coming under great scrutiny on the related topics of transparency in clinical trial process and transparency in the publication of results from clinical trials [3]. In 2004, GlaxoSmithKline was named in a lawsuit filed by the state of New York (USA) [4]. The suit alleged that GlaxoSmithKline had withheld the publication of crucial clinical data concerning the effects on children who received one of its drugs (the antidepressant paroxetine). Paroxetine had not received marketing approval in the United States for treating children and adolescents with depression, although it had marketing approval for its use in treating adults with depression. In the United States, physicians are allowed to prescribe any legal drug for any use, a practice termed 'off-label use'. Thus, while not labeled for use in children, children and adolescents were prescribed paroxetine by their physicians. The lawsuit alleged that GlaxoSmithKline had data from clinical trials, as early as 1998, suggesting that children and adolescents should not be treated with paroxetine for depression, but that the company had prevented the data from being published or otherwise made publicly available. The company settled the lawsuit by paying a substantial fine and by agreeing to establish a public, online database containing summaries of results of all its clinical trials, including the results of several trials of paroxetine in children and adolescents [5].

In September 2004, Merck & Company, Inc. stopped one of its ongoing clinical trials (Adenomatous Polyp Prevention on Vioxx, or the APPROVe Trial) and voluntarily removed the drug (rofecoxib) from the market [6]. The data from the trial indicated a statistically significant increase in confirmed thrombotic cardiovascular events, such as heart attacks and strokes, beginning after 18 months of treatment in patients taking the drug compared with patients taking placebo [7]. Much has been written about these events and the analysis of the previous VIGOR trial with rofecoxib [8–10]. Data from the VIGOR trial showed significant reductions in serious upper gastrointestinal adverse events in patients who took rofecoxib compared with patients who took naproxen; however, there was a statistically significant increase in thrombotic cardiovascular events, mainly heart attacks, in the patients who took rofexocib. Without a concurrent placebo-control group in the VIGOR study, it was unclear if the difference in cardiovascular events was due to an increase with rofecoxib, a decrease with naproxen, or a combination of the two.

A third controversy centered on celecoxib, a cyclooxygenase-2 (COX-2)-specific inhibitor marketed by Pfizer. Clinical research on celecoxib was originally designed to determine efficacy and the risk of gastrointestinal adverse effects, and different results were published concerning cardiovascular events. One study (Celecoxib Long-term Arthritis Safety Study, or the CLASS study) reported no increase in major cardiovascular events, including myocardial infarctions [11], while another study (Colorectal Adenoma Prevention Trial or the APC study), reported a 2.3- and a 3.4-fold increased risk of cardiovascular events with the 400-mg and 800-mg daily doses of celexcoxib, respectively [12]. Other papers analyzed data from clinical trials of COX-2 inhibitors in different ways and found no risk or an increased risk [13–18]. Relevant to the issue of clinical trial registration is the fact that the authors published only 6 months of data when CLASS was a 12- to act that prohibits pharmaceutical companies from advertising prescription drugs unless the company has disclosed to state health department officials certain information concerning the clinical trials of the drugs [29]. Under the new law, which became effective in October 2005, drug manufacturers are required to disclose the name of the group conducting the clinical trial, the purpose of the trial, dates during which the trial was conducted, and information about the results of the trial, including potential or adverse effects of the drug. Maine established a publicly accessible Web site that links to clinicaltrials.gov.

Other states have pending bills and legislation on the conduct of clinical trials and the dissemination of results.

Journal editor actions

In September 2001, 12 journals simultaneously published a paper by editors from the International Committee of Medical Journal Editors (ICMJE) calling on physicians and academic centers to be more involved in clinical trial research [3]. The ICMJE stated that their journals would not accept manuscripts unless the investigators had had meaningful involvement in all aspects of the study, including writing of the subsequent manuscript, and made the decision to publish the results.

In September 2004, the ICMJE announced that they would not consider for publication, the results of any trial that had not been prospectively registered in a publicly accessible database [30, 31]. The ICMJE stated that they would require, as a condition of consideration for publication, all new trials started after 1 July 2005 to be registered at inception and would require all ongoing trials to be registered by 13 September 2005. The editors of the British Journal of Medicine, a member journal of ICMJE, issued their own statement concerning clinical trial registries [32]. The difference between the two editorials is the issue of the recognition of the appropriate trial registries. Neither statement endorses a given registry, but the ICMJE requires use of the Web site operated by the National Library of Medicine (www.clinicaltrials.gov). The British Medical Journal cites this registry and another one (www.cursi.co.uk). The latter Web site is operated by Current Science Ltd, a British publishing group that manages BioMed Central, which provides immediate free access to peer-reviewed journals.

International pharmaceutical industry association actions

In response to the ICMJE September 2001 editorial and to other editorials, and a growing public dissatisfaction, the Pharmaceutical Research and Manufacturers Association (PhRMA) issued guidelines for the conduct of clinical trials and the publication of results [33]. PhRMA members committed to the timely communication of meaningful clinical trial results of marketed products or of investigational products approved for marketing, regardless of study outcome. The PhRMA guidelines allowed delays in publication for the purpose of protecting intellectual property.

In June 2005, PhRMA issued a news release stating that PhRMA member companies agreed to post their clinical studies on the National Library of Medicine Web site (www.clinicaltrials.gov) as of 1 July 2005 [34].

PhRMA, the European Federation of Pharmaceutical Industries and Associations, the International Federation of Pharmaceutical Manufactures and Associations, and the Japanese

Pharmaceutical Manufacturers Association joined together to agree on voluntary principles for disclosing information about clinical trials [35].

Discussion

The posting of clinical trial information and clinical trial results has the potential to make the process of drug development more transparent to the public. It also has the potential to make a drug sponsor's development plan more transparent to competitors. Drug sponsors may spend millions of dollars over many years refining their clinical development plans for a given therapeutic. With the clinical hypothesis, endpoints, and statistical design made public, astute competitors can design clinical trials with a competing product. It is possible for the second company to leap ahead of the original drug sponsor and obtain marketing approval first, or at a minimum, to shave substantial amounts of time off its development timeline. In this situation, innovation may be curtailed. On the other hand, it is possible that making some clinical trial information public will make clinical research more efficient by discouraging duplicative efforts.

It is not clear if clinical trial registry information in its present format is valuable to patients [36]. Patients may not be interested in endpoints or power of statistical analysis, and may only wish to learn of opportunities to try a new medication, particularly if other therapies have failed to help them. Journal editors can certainly tell if a paper based on a clinical protocol reports the endpoints that were prospectively defined by checking the clinical trial registry. Is it necessary that this information be open to public posting and public scrutiny? At the WHO meeting in Geneva, Switzerland (April 2005), it was agreed that certain fields (i.e., 5 of the 20 in the minimum data set) might be disclosed at a later time if the study sponsor believed that their early disclosure would cause competitive disadvantage. Members of PhRMA have agreed that they would be willing to supply the full protocol, statistical analysis plan, and any amendments to the journal at the time the paper was submitted; these items could satisfy the ICMJE concerns without jeopardizing the drug sponsor.

As for the dissemination of results, most drug sponsors hope that the data from their clinical trials will be published in a peer-reviewed journal. Traditionally, it has been difficult for studies with 'negative data' (i.e., study results did not support the hypothesis) to be published [37]. However, science does move forward knowing both positive and negative data, and dissemination of results for all clinical trials is warranted. It will be instructive to monitor the ICMJE journals to see whether they hold to the spirit of transparency by publishing 'negative' data. If negative data are not published on a more-frequent basis, then the format of clinical trial result dissemination remains a problem.

Some drug sponsors have decided to present all data on a clinical trial on their company Web sites, often in the format of a clinical study report of 30 or more pages. If the aim of dissemination of the results is to assist patients in learning about their disease or learning the outcome of the clinical trial in which they participated, it is not clear if the clinical study report format is the best format. Many patients are not familiar with clinical trial procedures, statistical analyses, or even medical terminology. It is possible that this format may provide too much information, in an inappropriate and unhelpful manner.

Much work remains to be done with both the registration of trials at inception and the dissemination of results in a timely manner. Patients who enroll in clinical trials or who are considering enrolling in clinical trials should have access to the basic plan of a study in the informed consent process and be informed of the results in at least a summary manner. Drug sponsors who finance the clinical trials should have protection for their proprietary information. A cooperative alliance among patients, sponsors, regulators, and editors will allow a common ground to be found.

Acknowledgements

I am grateful to Dr. Philip David Noguchi and Dr. Laurence Hirsch for their critical reviews.

References

1   Simes RJ (1986) Publication bias: the case for an international registry of clinical trials. *J Clin Oncol* 4: 1529–1541
2   Chalmers I (1990) Underreporting research is scientific misconduct. *JAMA* 263: 1405–1408
3   Davidoff F, DeAngelis CD, Drazen JM et al (2001) Sponsorship, authorship, and accountability. *N Engl J Med* 345: 825–826
4   Office of the New York State Attorney General Eliot Spitzer. Major pharmaceutical firm concealed information. Available at http://www.oag.state.ny.us/press/2004/jun/jun2b_04.html (Accessed 17 March 2005)
5   GlaxoSmithKline Clinical Trial register. Available at http;//ctr.gsk.co.uk/welcome.asp (Accessed 23 March 2005)
6   Merck. Merck announces voluntary worldwide withdrawal of VIOXX. September 30, 2004. Available at http://www.merck.co.vioxx_withdrawl/pdf/vioxx_press_release_final.pdf (Accessed 13 April 2006)
7   Gregory D, Curfman GD, Morrissey S, Drazen JM (2005) Expression of concern: Bombardier et al., "Comparison of upper gastrointestinal toxicity of rofecoxib and naproxen in patients with rheumatoid arthritis" *N Engl J Med* 353: 2813–2814
8   Bombardier C, Laine L, Reicin A et al (2000) Comparison of upper gastrointestinal toxicity of rofecoxib and naproxen in patients with rheumatoid arthritis. *N Engl J Med* 343: 1520–1528
9   Juni P, Nartey L, Reichenbach S, Sterchi R, Dieppe PA, Egger M (2004) Risk of cardiovascular events and rofecoxib: cumulative meta-analysis. *Lancet* 364: 2021–2029

10 Bombardier C, Laine L, Burgos-Vargas R et al (2006) Response to expression of concern regarding VIGOR study. *N Engl J Med* 354: 1196–1199

11 Silverstein FE, Faich G, Goldstein JL et al (2000) for the Celecoxib Long-Term Arthritis Safety Study. Gastrointestinal toxicity with celecoxib vs nonsteroidal anti-inflammatory drugs for osteoarthritis and rheumatoid arthritis: the CLASS study: a randomized controlled trial. *JAMA* 284: 1247–1255

12 Solomon SD, McMurray JJ, Pfeffer MA et al (2005) Adenoma Prevention with Celecoxib (APC) Study Investigators. Cardiovascular risk associated with celecoxib in a clinical trial for colorectal adenoma prevention. *N Engl J Med* 352: 1071–1080

13 Hrackovec JB, Mora M. Reporting of 6-month vs 12-month data in a clinical trial of celecoxib. *JAMA* 386: 2398

14 Mukherjee D, Nissen SE, Topol EJ. Risk of cardiovascular events associated with selective COX-2 inhibitors. *JAMA* 286: 954–959

15 Pitt B, Pepine C, Willerson JT. Cyclooxygenase-2 inhibition and cardiovascular events. *Circulation* 106: 167–169

16 Mamdani M, Rochon P, Juurlink DN et al (2003) Effect of selective cyclooxygenase-2 inhibitors and naproxen on short-term risk of acute myocardial infarction in the elderly. *Arch Intern Med* 163: 481–486

17 Solomon DH, Schneesweiss S, Glynn RJ et al (2004) Relationship between selective cyclooxygenase-2 inhibitors and acute myocardial infarction in older adults. *Circulation* 109: 2068–2073

18 Graham DJ, Campen D, Hui R et al (2005) Risk of acute myocardial infarction and sudden cardiac death in patients treated with cyclooxygenase 2 selective and non-selective non-steroidal anti-inflammatory drugs: nested case-control study. *Lancet* 365: 475–481

19 Eckert CH. Bioequivalence of levothyroxine preparations: industry sponsorship and academic freedom. *JAMA* 277: 1200–1201

20 Rennie D (1997) Thyroid storm. *JAMA* 277: 1238–1243

21 Spigelman MK. Bioequivalence of levothyroxine preparations for treatment of hypothyroidism. *JAMA* 277: 1199–1200

22 Weiss RB, Rifkin RM, Stewart FM et al (2000) High-dose chemotherapy for high-risk primary breast cancer: an on-site review of the Bezwoda study. *Lancet* 355: 999–1003

23 Weiss RB, Gill CG, Hudis CA (2001) An on-site audit of the South African trial of high-dose chemotherapy for metastatic breast cancer and associated publications. *J Clin Oncol* 19: 2771–2777

24 European Commission. Enterprise an Industry Directorate-General. Guideline on the data fields from the European clinical trials database (EudraCT) that may be included in the European data base on Medicinal Products. Available at: http://pharmacos.eudra.org/F2/pharmacos/docs/

Why did ICMJE call for mandatory clinical trial registration?

Calls for trial registration are not new [5, 6]. Registration of clinical trials has been seen as the best measure to reduce publication bias and, hopefully, to insure the integrity of results presented to the public. Several cases of scientific and ethical failures in published studies, however, increased the concerns of journal editors about bias, and prompted ICMJE to take concrete measures to improve the quality of the studies published in their journals. For example, the Journal of the American Medical Association published results of an important study of celecoxib, one of the inhibitors of cyclooxygenase (COX)-2 [7], for which it was later claimed that only 6-month data were reported, although 12-month data were available [8]. The New England Journal of Medicine published its concern about the article on another COX-2 inhibitor, rofecoxib [9] because data showing the cardiovascular effects at a later stage of the study were not included in the published report [10]. The journal's editors believed that the authors were not aware of the three cases that, if included, would have significantly changed the conclusion of the study. Material obtained by subpoena in the Vioxx (trade name for celecoxib) litigation and made available to the journal showed that at least two authors of the published study knew about the additional cases before the first revision of the submitted article [10].

How did ICMJE develop a trial registration policy?

The COX-2 papers and other examples of scientific and ethical problems in reporting results from clinical trials [11] reached a point at which many journal editors felt that they had the responsibility to do what they could to protect the integrity of published material, and to maintain the trust of the professional and lay public in the articles the journals published. The best solution, from the point of evidence-based medicine, would be complete transparency and public access to all aspects of clinical research, from trial registration to publishing, not necessarily in journals, of all data from clinical trials. Complete transparency, however, also has problems. Journal editors were aware that, at that time, no immediate or complete solution for these problems was in place, and so decided to address individual problems one at a time. The journal editors also thought that by making decisions one at a time, all stakeholders in clinical research, from patients to researchers and pharmaceutical or device manufacturers, would have time and evidence to assess needs, develop principles, discuss problems, and agree on implementation actions.

At the annual meeting of the ICMJE in Dubrovnik, Croatia, in 2004, the editors decided to begin with a call for comprehensive registration of the trials themselves (and not, for instance, the results of the trials) as the first step in alleviating selective data presentation on clinical trials in medical literature. Eleven ICMJE member journals adopted a trials-registration policy in which they required registration of a trial in a public registry as a condition of consideration for publication. In the editors' first editorial, published in September 2004 [1], they delineated that the registration of the trial should occur at or before the onset of patient enrollment. They also defined the type of trial that should be registered, namely, any research project that prospectively assigns human subjects to intervention or comparison groups to study the cause-and-effect relationship between a medical intervention and a health outcome [1]. This definition does not include studies for other purposes, such as phase 1 trials. The ICMJE editors provided ample time for researchers to learn about this publishing requirement, set-

ting 1 July 2005 as the date after which the policy would apply
to any new clinical trial. For trials that began enrollment before
this date, 13 September 2005 was the deadline for registration.

In the first editorial, the editors defined the minimum information needed to be registered: a unique identifying number, a statement of the intervention (or interventions) and
comparison (or comparisons) studied, a statement of the study
hypothesis, definitions of the primary and secondary outcome
measures, eligibility criteria, key trial dates (registration date,
anticipated or actual start date, anticipated or actual date of
last follow-up, planned or actual date of closure to data entry, and date trial data considered complete), target number of
subjects, funding source, and contact information for the principal investigator [1].

In April 2005, some members of the ICMJE participated in a
meeting organized by the World Health Organization (WHO),
which brought together all stakeholders in clinical trials, including representatives from the pharmaceutical companies, patients, journal editors, researchers, and government representatives involved in national trial registries. At the meeting, it was
agreed that a minimal data set of 20 items must be provided
(Tab. 1). At the ICMJE annual meeting in Bethesda, Maryland,
USA, in May 2005, the ICMJE also adopted the WHO minimal data set as the ICMJE's requirement. In the second IC-
MJE editorial statement [2], entering data for all 20 fields from
the WHO minimal data set was required for consideration of a
submitted manuscript presenting the results of a clinical trial.

The future clinical trials and medical journals

With trial registration as a part of current requirements for publication of results in many journals, both the researchers and

Table 1. Minimal data set for trial registration defined by the World Health Organization and adopted by the International Committee of Medical Journal Editors for their registration policy

Item		Explanation
1.	Unique trial number	The unique trial number will be established by the primary registering entity (the registry)
2.	Trial registration date	The date of registration will be established by the primary registering entity
3.	Secondary IDs	May be assigned by sponsors or other interested parties (there may be none)
4.	Funding source(s)	Name of the organization(s) that provided funding for the study
5.	Primary sponsor	The main entity responsible for performing the research
6.	Secondary sponsor(s)	The secondary entities, if any, responsible for performing the research
7.	Responsible contact person	Public contact person for the trial, for patients interested in participating
8.	Research contact person	Person to contact for scientific inquiries about the trial.
9.	Title of the study	Brief title chosen by the research group (can be omitted if the researchers wish)
10.	Official scientific title of the study	This title must include the name of the intervention, the condition being studied, and the outcome (e.g., The International Study of Digoxin and Death from Congestive Heart Failure).
11.	Research ethics review	Has the study at the time of registration received appropriate ethics committee approval (yes/no)? (It is assumed that all registered trials will be approved by an ethics board before commencing)
12.	Condition	The medical condition being studied (e.g., asthma, myocardial infarction, depression)
13.	Intervention(s)	A description of the study and comparison/control intervention(s) (for a drug or other product registered for public sale anywhere in the world, the generic name should be given; for an unregistered drug the generic name or company serial number is acceptable). The duration of the intervention(s) must be specified.

Table 1. (continued)

Item		Explanation
14.	Key inclusion and exclusion criteria	Key patient characteristics that determine eligibility for participation in the study
15.	Study type	Database should provide drop-down lists for selection, including choices for randomized *versus* non-randomized, type of masking (e.g., double-blind, single-blind), type of controls (e.g., placebo, active), and group assignment (e.g., parallel, crossover, factorial)
16.	Anticipated trial start date	Estimated enrollment date of the first participant
17.	Target sample size	The total number of subjects the investigators plan to enroll before closing the trial to new participants
18.	Recruitment status	Is this information available (yes/no) (if yes, link to information)
19.	Primary outcome	The primary outcome that the study was designed to evaluate. Description should include the time at which the outcome is measured (e.g., blood pressure at 12 months)
20.	Key secondary outcomes	The secondary outcomes specified in the protocol. Description should include time of measurement (e.g., creatinine clearance at 6 months).

their sponsors want clear answers to the several questions:
- Which trial should be registered to allow the publication of the results in a medical journal?
- Which registration database should be used?
- Which data must be registered?
- Which journals subscribe to the ICMJE trial registration policy?

Which trials should be registered?

As explained in the two editorials [1, 2], ICMJE wants to ensure public access to all clinical trials that test any clinical hypothesis about health intervention and its outcomes – both before and after publication of the trial. Examples of such trials are the testing a new drug or medical device against the current standard. Trials with the primary goal of assessing drug toxicity or determining pharmacokinetics of a drug are not required to be registered (phase 1 trials). Between the phase 1 trials, which do not need to be registered, and the phase 3 trials, which must be registered because they provide clinically directive answers, there may be clinical trials for which it may be difficult to decide when to register. Registration of trials such as those investigating the biology of disease or those providing preliminary data that may lead to larger, clinically directive trials could possibly slow the innovation drive. ICMJE policy leaves the decision on registration to each individual journal and case-by-case review of trials in this category. The best answer to doubts about a trial is to register it. It is important to emphasize that registration of trials is not mandatory for doing clinical research. Only if the researchers and their sponsors want to use medical journals as the medium for communicating the results to the public is registration of clinically directive trials mandatory. Journal editors firmly believe that this policy is the best way to repay the altruism of volunteers who risk their health and lives in clinical trials.

Which registration database can be used?

The ICMJE does not advocate any particular registry or registries, and any registry that meets the criteria is acceptable. The criteria are that the registry:
- must be accessible to the public at no charge
- must be open to all prospective registrants (meaning that investigators are able to register without restriction by geographic location, academic affiliation, patient demographics, or clinical condition)

- must be managed by a not-for-profit organization
- must have be a mechanism to ensure the validity of the registration data
- should be electronically searchable
- must include all data from the minimal data set (Tab. 1).

To date, only a few databases satisfy these criteria (Tab. 2) and are listed at the ICMJE Web site (www.icmje.org). A number of national and international trial registries are under development [12]. The editors recognize that too many registries may be confusing both to the researchers and journal editors, and the editors look forward to further development of the recently launched International Clinical Trials Registry Platform (ICTRP) of the WHO, which aims to set international norms and standards for trial registration and reporting [13].

A database that is electronically searchable and is logically organized is important, as it allows the editors to quickly retrieve the information on the minimal data set. The identifying number of the trial should be stated in the submitted manuscript, best at the end of the abstract, where it will be published.

What data should be entered in the registration database?
The ICMJE requires the whole data set as defined in its editorial [2] (Tab. 2). Inclusion of some of the items in the data set has been debated, with concerns that some of the information may be proprietary and perhaps should not be disclosed to the public until the publication of the study, but made available only to the journal editor evaluating the submitted manuscript [14]. Although different stakeholders in clinical trails have different view about what information should be disclosed in the database, editors believe that full disclosure of the data is necessary. Moreover, it is important to complete the data-

Table 2. Trial regisration databases which satisfy ICMJE requirements

Web-site of the registry	Description
http://www.clinicaltrials.gov	A service of the US National Institutes of Health, developed by the National Library of Medicine. Provides regularly updated information about federally and privately supported clinical research in human volunteers
http://isrctn.org	Online service that provides unique numbers to randomized controlled trials in all areas of healthcare and from all countries around the world
http://www.actr.org.au	The Australian Clinical Trials Registry (ACTR) is a national, online register of clinical trials being undertaken in Australia
http://www.umin.ac.jp	University Hospital Medical Information Network (UMIN) clinical trial registry in Japan
http://www.trialregister.nl/trialreg/index.asp	Dutch trial register, run by the Dutch Cochrane Center

base with meaningful data. For example, a recent analysis of registered trials at www.Clinicaltrials.gov showed that almost 25% of the registrations from studies sponsored by industry had left blank the field for the "primary outcome measure" [4]. Of the registrations that had completed this field, 17% of the entries were vague, and 19% did not specify the measure. Such results from the largest registration database are good arguments for full disclosure of the minimum data set.

Which journals subscribe to the trial registration policy?
In addition to the 11 journal members of the ICMJE, a number of journals have published the ICMJE statements or explicitly defined their requirements for trial registration. Table 3 lists such journals indexed in the PubMed bibliographic database. This list will certainly enlarge with time. Authors should check

Table 3. List of medical journals who have published the ICMJE statements or adopted the ICMJE policy, including the 11 journal members of the ICMJE*

Journal members of the ICMJE:
British Medical Journal
Canadian Medical Association Journal
Croatian Medical Journal
Danish Medical Journal
Dutch Medical Journal
JAMA
Lancet
Medical Journal of Australia
New England Journal of Medicine
New Zealand Medical Journal
The Journal of the Norwegian Medical Association

Journals that published or commented on the ICMJE statement on the registration policy:
AIDS (2005) 19(2): 105
Am J Transplant (2005) 5(4 Pt 1): 643
American Journal of Physical Medicine & Rehabilitation (2005) 84(1): 3–4
Arch Dis Child (2006) 91(1): 93
Archives of General Psychiatry (2006) 63(1): 100
Archives of Ophthalmolology (2005) 123(9): 1263-4
Archives of Dermatology (2005) 141(1): 76–77, discussion 75
Archives of Otolaryngology – Head & Neck Surgery (2005) 131(6): 479–480
Arteriosclerosis, Thrombosis & Vascular Biology (2005) 25(4): 873–874
Arthritis and Rheumatism (2005) 52(8): 2243–2247
British Journal of Dermatology (2005) 152(5): 859–860
Circulation Research (2005) 96(5): 600–601
Circulation (2005) 111(10): 1337–1338
Clinical Trials (2005) 2(2): 193
Contemporary Clinical Trials (2005) 26(5): 517
Hypertension (2005) 45(4): 631–632
Indian Journal of Medical Ethics (2005) 2(3): 74–75
Investigacion Clinica (2004) 45(4): 295–296
Journal of the American Academy of Dermatology (2005) 52(5): 890–892
Journal of the American Osteopathic Association (2004) 104(10): 409–410
Journal of the American Sociology of Nephrology (2005) 16(4): 837
Journal of Medical Internet Research (2004) 6(3): e35
Journal of National Cancer Institute (2005) 97(6): 410–411
Journal of Athletic Training (2005) 40(1): 8
Medicina Clinica (2005) 124(16): 638–639
Nephrology, Dialysis, Transplantation (2005) 20(4): 691
Paediatric Anaesthesia (2006) 16(1): 92
PLoS Med (2005) 2(11): e365

Table 3. (continued).

Report on Medical Guidelines & Outcomes Research (2004) 15(19): 1–2, 6–7
Stroke (2005) 36(4): 924–925
Transplantation (2005) 79(7): 751

According to search of PubMed and PubMed Central using the combination of the key words "trial registration", "International Committee of Medical Journal Editors" (or "ICMJE"); search performed on 1 February 2006.

a journal's guidelines for authors to determine if the journal subscribes to the trial registration policy.

Some medical journal editors from small and developing countries are concerned that mandatory registration may influence both the research and medical journals in these settings [15]. Researchers in developing countries face many obstacles in their research and may not be aware or be able to easily comply with the registration process. Editors of journals from developing countries already have a shortage of good submissions, and registration requirement may further worsen their existence.

Another concern is that English as the language of registration may cause problems because the concepts and definitions of the data elements that must be registered do not translate directly into other languages. This problem must be solved by accompanying the data fields by more detailed explanations in different languages, by creating Web interfaces of the registration databases in different languages, or both, while keeping the database itself in English.

Discussion

The ICMJE editorial policy of trial registration may prove to be an opportunity for editors in small and developing communities to inform their authors and readers about current requirements for publishing clinical research and to promote the practice of trial registration. Journal editors, as respected professionals and researchers in their own community and at the same time most knowledgeable about the newest developments in the publishing area, are instrumental for increasing awareness and understanding of the need for trial registration, not only for their authors and readers, but also the wider community [16]. The editors can actively promote registration by alerting institutional review boards or ethical committees at hospitals and academic and research institutions. Because all research with human subjects must be approved by such legal bodies, information about trial registration would then reach all interested parties at the stage of trial planning and not at the stage of manuscript submission, when it is too late.

We do not think that that adherence to registration policy will decrease the number of articles in small journals. Firstly, clinical trials in smaller and developing communities are often a part of larger multicenter trials or are conducted by large pharmaceutical companies. These companies should be well aware of the ICMJE registration policy, as they are most interested in publishing the results about their new pharmaceutical products or devices. In fact, researchers from the smaller research communities should make sure that the pharmaceutical firm approaching them about a possible trial has registered, or plans to register, it before the enrollment of the first participant. Secondly, we believe that trial registration can benefit independent research from small and developing countries because registration would make such studies visible to the global scientific community, and thus contribute to the protection of their intellectual rights. Small countries may have more problems in choosing where to register, as their national, regional, or disease-specific trial registration databases [12] may not satisfy all ICMJE requirements.

Publication of trial results is the public presentation of the research. Medical journals are the main medium for public presentation of clinical trials, and their mandatory registration ensures that editors can check the results in a submitted manuscript against the proposed trial protocol, and thus ensure that the presentation is honest, transparent, and accurate. Journal editors will closely follow and analyze the implementation of trial registration policy. The analysis of the largest registration database [4] is the first example of such follow-up: it identified some problems and suggested possible solutions, and it is hoped resolved some disagreements and challenges. As stated in the second ICMJE editorial [2], the registration policy will be maintained for the next 2 years and then reviewed. We are entering a very challenging but equally stimulating and exciting period, and all stakeholders should be involved in gathering evidence on how this policy is working.

References

1   De Angelis C, Drazen JM, Frizelle FA et al. Clinical trial registration: a statement from the International Committee of Medical Journal Editors. Available at www.icmje.org (Accessed: 29 January 2006)
2   De Angelis C, Drazen JM, Frizelle FA et al. Is this clinical trial fully registered? A statement from the International Committee of Medical Journal Editors. Available at: www.icmje.org (Accessed: 29 January 2006)
3   Drazen JM, Wood AJJ (2005) Trial registration report card. *N Engl J Med* 353: 2809–2811
4   Zarin DA, Tse T, Ide NC (2005) Trial registration at ClinicalTrials.gov between May and October 2005. *N Engl J Med* 353: 2779–2787
5   Simes RJ (1986) Publication bias: the case for an international registry of clinical trials. *J Clin Oncol* 4: 1529–1541
6   Dickersin K, Rennie D (2003) Registering clinical trials. *JAMA* 290: 516–523

7   Silverstein FE, Faich G, Goldstein JL et al (2000) Gastrointestinal toxicity
    with celecoxib vs nonsteroidal anti-inflammatory drugs for osteoarthritis
    and rheumatoid arthritis: the CLASS study: a randomized controlled trial.
    Celecoxib Long-Term Arthritis Safety Study. *JAMA* 284: 1247–1255
8   Hrachovec JB, Mora M (2001) Reporting of 6-month vs 12-month data in
    a clinical trial of celecoxib. *JAMA* 386: 2398
9   Bombardier C, Laine L, Reicin A et al (2000) Comparison of upper gastro-
    intestinal toxicity of rofecoxib and naproxen in patients with rheumatoid
    arthritis. *N Engl J Med* 343: 1520–1528
10 Gregory D, Curfman GD, Morrissey S, Drazen JM (2005) Expression of
    concern: Bombardier et al., "Comparison of upper gastrointestinal toxicity
    of rofecoxib and naproxen in patients with rheumatoid arthritis" N Engl J
    Med 2000; 343:1520–1528. *N Engl J Med* 353: 2813–2814
11 Fontanorosa PB, DeAngelis CD (2005) Conflict of interest and indepen-
    dent data analysis in industry-funded studies. *JAMA* 294: 2576
12 Haug C, Gøtzsche PC, Schroeder TV (2005) Registries and registration of
    clinical trials. *N Engl J Med* 353: 2811–2812
13 World Health Organization. WHO International Clinical Trials Registra-
    tion Platform: Unique ID assignment. Geneva: WHO. Available at http://
    www.who.int/ictrp/commnets2/en/index.html (Accessed 23 January 2006.)
14 Krleža-Jerić K (2005) Clinical trial registration: the differing views of in-
    dustry, the WHO, and the Ottawa group. *PLoS Medicine* 2: e387
15 Habibzadeh F (2006) Call for mandatory registration of clinical trials and
    its impact on small medical journals: a scenario of an emerging bias. *Croat
    Med J* 47: 181-182
16 Marušić M, Marušić A (2001) Good editorial practice: editors as educators.
    Croat Med J 42: 113–120

Industry perspective on public clinical trial registries and results databases

Tracy Beck

Eli Lilly and Company, Lilly Corporate Center, Indianapolis, IN 46285, USA

Introduction

Recent initiatives among universities, government agencies, and biopharmaceutical/biotech companies have focused on the disclosure of information on clinical trials to increase transparency. This issue is not new: in 1997, the FDA Modernization Act (FDAMA) (Section 113) directed the Secretary of Health and Human Services to establish, maintain, and operate a databank of information on clinical trials for drugs to treat serious or life-threatening diseases and conditions conducted under the FDA's investigational new drug (IND) regulations (21 CFR part 312) [1], and in 2000, the National Institutes of Health (NIH) launched clinicaltrials.gov. In 2002, the Pharmaceutical Research and Manufacturers of America (PhRMA) issued its Principles on Conduct of Clinical Trials and Communication of Clinical Trial Results [2], and in June 2004, PhRMA sponsored and launched clinicalstudyresults. org, a database for clinical trial results [3]. While increasing transparency of clinical trials has the support of the member organizations of the European Federation of Pharmaceutical Industries and Associations (EFPIA), the International Federation of Pharmaceutical Manufacturers and Associations (IFPMA), the Japanese Pharmaceutical Manufacturers

Association (JPMA), and PhRMA, it will be important for continued innovation that increased transparency does not compromise intellectual property. A balance between transparency and intellectual property is the best way to ensure a continuation of innovative treatments for patients. To achieve a balance, decisions regarding disclosure requirements should include industry representation.

This chapter summarizes information from discussions with several biopharmaceutical companies and is not necessarily the perspective of any one company. The chapter addresses various perspectives on the issues and options in the current debate regarding registering clinical trials at study initiation and disclosing the results of these trials at completion. The debate is ongoing and the process is evolving.

Clinical trial registries

According to the PhRMA, the intent of a clinical trial registry is to provide patients and healthcare providers access to information about ongoing and enrolling clinical trials, and to provide a public record and status of those clinical trials [4]. Numerous types of publicly accessible registries are currently available for sponsors of clinical trials to provide information about initiated and ongoing clinical trials.

Non-industry-sponsored sites also exist, such as clinicaltrials.gov, which is operated by the NIH through the National Library of Medicine (NLM). The original purpose of clinicaltrials.gov was to provide a registration site for federally and privately funded clinical trials for experimental treatments (drug and biological products) for patients with serious and life-threatening diseases, allowing patients an opportunity to enroll in clinical trials and access to investigational therapies that would not be known to them. The information available to patients on the website was: eligibility criteria for participation in the trial, a description of the location of trial sites, and a point of contact for those wanting to enroll in the trial [1]. The original intent of the Web site shifted with the International Committee of Medical Journal Editors (ICMJE) position, when they announced that they would not publish results of any clinical trial that had not been registered in a publicly accessible Web site before the trial started [5]. Although ICMJE did not advocate a particular registry, the ICMJE established specific requirements, which are currently met by the Web site, clinicaltrials.gov.

The industry is supportive of registering clinical trials in addition to those for serious and life-threatening diseases, not only for transparency, but also to allow patients the opportunity to participate in clinical trials. Although not fully evaluated, registries may provide a vehicle to enhance recruitment of potential subjects into clinical trials, thereby reducing the overall time to complete a study.

The National Cancer Institute (NCI) sponsors a Web site, cancer.gov/clinicaltrials. This site contains information on cancer clinical trials that are open/active and accepting patient enrollment, including trials for cancer treatment, genetics, diagnosis, supportive care, screening, and prevention. The site also provides access to information on closed clinical trials that have been completed or are no longer accepting patients. In addition, NCI sponsors a Physician Data Query (PDQ) site that contains peer-reviewed summaries on cancer treatments, screening, prevention, genetics, and supportive care, and complementary and alternative medicine [6]. A number of biopharmaceutical companies also sponsor their own Web sites for disclosing clinical trial information at the inception of the trial, as well as results upon completion. Given the number of different registries available, and the ongoing proliferation of new registries, the IFPMA launched a Web portal in 2005, ifpma.org/clinicaltrials.html, which makes it easy for patients and providers to locate clinical trial registries and databases in any region by providing links to global clinical trial information [7]. Currently, this portal links to nine sites, with plans in place to support additional linkages.

In January 2005, the organizations representing global pharmaceutical companies (EFPIA, IFPMA, JPMA, and PhRMA) issued a Joint Position on the Disclosure of Clinical Trial Information via Clinical Trial Registries and Databases [8]. The position stated that the pharmaceutical industry was committed to submitting all clinical trials, other than exploratory, for listing in a publicly accessible clinical trial registry within 21 days of the initiation of patient enrollment, unless there were alternative national requirements. In addition, the position stated that the information included would include: at minimum a brief title; trial description in lay terminology; trial phase; trial type (e.g., interventional); trial status; trial purpose (e.g., treatment, diagnosis, prevention); intervention type (e.g., drug, vaccine); condition or disease; key eligibility criteria, including age and sex; the location of the trial; and contact information.

The World Health Organization (WHO) Technical Consultation on Clinical Trial Registration Standards meeting in April 2005 outlined which trials would be registered (any research that prospectively assigns human volunteers to one or more health-related interventions to evaluate the effects on health outcomes), the minimum amount of information that would be registered (Tab. 1), standards for disclosing results of completed trials (the non-promotional ICH E3 template), and the need to convene a group to develop a mechanism to delay release (delayed disclosure mechanism) of one or more

Table 1. Comparison of minimum data set requirements for registration of clinical trials [9]

Minimum data set requirements		WHO/PhRMA/ IFPMA/JPMA/ EFPIA	ICMJE
1.	Unique trial number	X	X
2.	Registration date	X	X
3.	Secondary IDs	X	X
4.	Funding source(s)	X	X
5.	Primary sponsor	X	X
6.	Secondary sponsor	X	X
7.	Responsible contact person	X	X
8.	Research contact person	X	X
9.	Brief title of study	X	X
10.	Official scientific title*	X*	X
11.	IRB review	X	X
12.	Condition	X	X
13.	Intervention(s)/comparator*	X*	X
14.	Key inclusion/exclusion criteria	X	X
15.	Study type	X	X
16.	Start date	X	X
17.	Target sample size*	X*	X
18.	Recruitment status	X	X
19.	Primary outcome*	X*	X
20.	Key secondary outcome*	X*	X

*According to the WHO Technical Consultation of Clinical Trials Registration Standards, one or more of starred items may be regarded as sensitive for competitive reasons by the sponsor who may wish to delay release of the information.
EFPIA: European Federation of Pharmaceutical Industries and Associations; ICMJE: International Committee of Medical Journal Editors; IFPMA: International Federation of Pharmaceutical Manufacturers and Associations; JPMA: Japanese Pharmaceutical Manufacturers Association; PhRMA: Pharmaceutical Research and Manufacturers of America; WHO: World Health Organization

of the five sensitive data items until a specified date [9]. In September 2005, the joint position of IFPMA, EFPIA, JPMA, and PhRMA was updated to reflect their position on the disclosure of sensitive information on public clinical trial registries, with industry supporting the use of the minimum data set identified during the WHO Technical Consultation on Clinical Trials Registration Standards meeting [10]. The global industry agreed on a minimum registry data set of 20 items to be disclosed at the inception (within 21 days of first patient visit) of hypothesis-testing/confirmatory clinical trials, regardless of disease state, with the caveat that, in infrequent situations, the sponsor may elect to delay the release of one, or more, of five items regarded as sensitive for competitive reasons by the sponsor (Tab. 2).

The ICMJE issued a policy in September 2004, requiring registration of clinical trials at inception before the results will be considered for publication by their member journals [5]. Additional non-member publishers have stated that they plan to accept this position. The ICMJE scope was further defined in May 2005, when they clarified that clinical trials were those that influence the healthcare decision-making process [11]. The ICMJE also endorsed the minimum data set of 20 fields recommended by the WHO, without the caveat that in certain situations a sponsor may elect to delay disclosure of the five select fields.

Overall, the concept of registering a clinical trial at inception has the support of the biopharmaceutical industry. Although the disclosure of sensitive information in the five fields is controversial, a further concern is disclosing sensitive data for all studies being conducted by a company in a single location (i.e., on a single database). Some proponents of complete disclosure claim that the data on which a company wants to delay public disclosure, are readily available, which

Table 2. Five items that may be considered sensitive by biopharmaceutical companies

- Official scientific title
- Intervention/comparator
- Target sample size
- Primary outcome
- Key secondary outcome

may be true, but in a very diverse and disseminated fashion. That is, while patients may compare notes about clinical trials in online chat rooms, and while journalists or analysts may post speculative articles about early clinical development of compounds on the Internet outside an online registry, no single public location contains information on all clinical trials in a therapeutic area from a particular company. The biopharmaceutical industry remains concerned not only with the timing of the disclosure of sensitive data at the individual trial level, but also that the disclosure of this sensitive data on the whole portfolio presents an even greater threat. Information on trial design, timing, and endpoints describes the specifics of the company's entire product development plan. Public disclosure can provide a competing company with a competitive advantage. Significant investments in time and money made by the sponsor in terms of creating an eloquent study design involving cutting-edge surrogate endpoints negotiated with regulators may become public simply by registering the comprehensive scientific title of the protocol. Competitors developing similar compounds could potentially reap this benefit, saving time and dollars. Given the time (10–15 years), effort, and cost of discovering and developing new drugs (estimated US $ 0.8–1.7 billion) [12, 13], losing a competitive advantage places innovation at risk.

Part of the concern with disclosing the data sets is related to the lack of agreement among industry and others regarding the scope of trials to be registered. The main issue is whether all trials (including early phase, exploratory, non-hypothesis testing trials, such as Phase 1) on all compounds, regardless of marketing status, require registration and disclosure at trial initiation. An "all-or-none" approach to disclosing trial data at initiation and completion could have consequences for sponsors who have invested substantial time and money in the discovery and development of new drugs. Early disclosure of clinical trial data to competitors would result in the loss of a competitive advantage, profits, and incentive to continue the costly drug development process. Posting the full clinical plan at trial inception, when the drug may not be on the market for over 5 years, is not beneficial or critical to patients or healthcare providers, but it is to the sponsor's business model. Posting all information requested by ICMJE for all trials at inception and posting results for all trials at completion could result in a decreased availability of new drugs for consumers, which could have a negative consequence on the health and well-being of the public.

In the absence of consensus on a delayed disclosure mechanism, some companies have implemented their own version of it. There is general agreement among most biopharmaceutical companies that the concern related to disclosing sensitive data occurs in the early phase exploratory trials (e.g., phase 1 and phase 2) and that delayed disclosure would only be requested in a few cases. One possible delayed disclosure method would be a fixed stage approach, whereby all 15 fields would be disclosed at inception for phase 2 trials (with the remaining commercially sensitive fields placed into a delayed disclosure mechanism). All 20 fields would be disclosed at inception for phase 3 trials, and once phase 3 trials begin, the 5 fields that were not disclosed from the phase 2 trials would be posted retrospectively.

Another possible method would be to use a third party to hold the sensitive data and allow disclosure as needed. With this method, the data could be date stamped at inception, electronically recorded, and provided to a third party. These data could be either disclosed or verification of its existence could be provided to appropriate individuals, such as journal editors, with approval from the sponsor. Such a process would allow the data to be captured for transparency with key audiences without jeopardizing the sponsor's intellectual property. Further exploration of electronic technology would likely reveal other potential secure data solutions.

If the intent of a registry is to allow patients and healthcare providers an opportunity to identify possible trials in which to enroll, and to provide transparency that trials are being disclosed so that sponsors will be accountable for the timely and accurate disclosure of results, there has yet to be a compelling argument as to how requiring additional data fields at inception fulfills that intent or improves public safety. Many people would argue that ultimately, it is the results of the clinical trial that are important, rather than specific information about how the study will be conducted. It is unclear, for example, how the sponsor's posting of the ICMJE's required 20 fields at inception would make any difference if adverse side effects occurred after a prespecified trial cut-off date. Clinical trials have a predetermined beginning and an end. Data reported in manuscripts and summaries are based on what was learned between these endpoints. Adverse side effects that occur after that period are reported to regulatory agencies. Journal editors typically do not want to publish positive results that may have occurred after that endpoint (because it occurred outside the window of the trial and therefore could introduce possible study bias); however, the opposite appears to be true if the results were negative.

Two surveys were conducted on public perceptions and relevant information desired on clinical trial registries. The results on public perceptions of clinical research studies, conducted by a Harris Interactive survey online from 19 to 26 April 2005 among a cross-section of 2261 adults in the United States aged 18 years and older, showed that 20% of adults indicated that they would be not at all likely to use an online clinical trial registry [14]. The participants in the survey reported that the media is their main source of receiving information about clinical research studies, although 51% of all adults indicated they would "prefer to learn about clinical research studies from their regular physician" [14]. There was no consensus on the purpose of registries, although most responses centered on information sharing: 40% indicated their purpose is to promote research, 36% said it is to increase awareness of clinical trials, and 32% said it is to share experiences of clinical trials. Of the respondents, 73% thought that all results should be available in these registries. It was concluded that "The public appears skeptical to the initial introduction of online clinical trial registries for themselves, but they are open to the information sharing they could provide" [14]. A survey on what patients want in a registry was conducted by the Center for Information and Study on Clinical Research Participation (CSICRP), a nonprofit organization dedicated to educating and informing the public, patients, medical/research communities, the media, and policy makers about clinical research participation [15]. The survey results provided insight into the information of concern to potential study subjects and survey participants (Tab. 3).

ICMJE has stated that they would refuse to publish a manuscript from any study that was not first registered at inception. Space constraints make it unlikely that medical journals will publish the results of all trials that have been registered. As

Table 3. *Survey results illustrate what concerns potential study subjects and survey participants [15]*

The information desired by most patients interested in enrolling in a clinical trial:
- description of trial
- trials seeking participants
- location of research center
- reputation of investigators
- glossary of terms
- compensation amount for participation

The information that was most important to the survey participants:
- geographic location of trial
- site reputation
- specific information about trials seeking volunteers
- investigator reputation
- compensation for participation
- funding sponsor
- number of study visits required

---

a result, journal editors must make judgments on which trial manuscripts will be published. Some concerns that critics have with the dissemination of results are access and timing. Many journals have a policy of not allowing predisclosure of data if a manuscript has been submitted to them for review. The review cycle for some journals can take months, with additional time between acceptance and the manuscript appearing in print. In addition, once the manuscript is in print, access may be limited to subscribers of the journal. Industry agrees that results databases can fill the gap by posting results of trials not published by journals, though there is concern that timing restrictions may compromise publication in the top-tiered journals. Indeed, some biopharmaceutical companies have strict policies that require public disclosure of trial results within 1 year of study completion, which could compromise the ability to publish those results in a top-tier journal. Because the medical community typically uses these peer-reviewed publications to obtain current clinical information, it is clear that the manu-scripts published in peer-reviewed journals and data posted in results databases should compliment each other, not replace one or the other. Peer-reviewed research is a cornerstone of evidence-based medical practice.

In addition to meeting the ICMJE registering requirements for publications, a number of Institutional Review Boards (IRB) have recently stated that they would not approve a clinical trial until it has been registered on clinicaltrials.gov and discloses all 20 fields. However, since a study cannot be registered as a clinical trial (available for patient enrollment) until at least one IRB has approved it, an apparent vicious cycle is formed, and it is unknown how companies will be able to comply.

In 2004, at least two legislative bills entitled the Fair Access to Clinical Trials (FACT) Act were introduced at the federal level in the United States [16, 17]. The proposed legislation would make it mandatory for biopharmaceutical companies to post information from their clinical trials on a public website at trial inception and completion. This type of legislation is also occurring at the state level and is cause for concern. Without a common standard for managing registry-related issues, the administrative burdens for a company to comply with differ-ent requirements from state to state could potentially result in delays in drug development, added costs, and the proliferation of registries.

Clinical trial results databases

Information on clinical trial results can be publicly disclosed on a number of different Web sites. Clinical trial results data-bases are public postings of the results of clinical trials, whether published in a medical journal or not, including negative findings and potentially adverse side effects [18]. The intent of clinical trial results databases is to enhance timely disclosure of clinical trial results (positive, negative, neutral, failed) to healthcare providers, to enable them to make informed, evidence-based treatment decisions. Because the results from all trials are not necessarily published in peer-reviewed journals, as some may be presented at scientific meetings, databases allow for the public disclosure of clinical trial results that might not otherwise be accepted and published by journals. If results are negative or neutral, they can be deemed by medical journals as not advancing medical debate and are, therefore, not newsworthy. In addition, due to space limitations, most peer-reviewed manuscripts do not report the outcome of every primary and secondary endpoint researched in the trial. The results databases can facilitate clinical trial transparency by ensuring that either the results are presented through a peer-reviewed reference citation or summarized according to the ICH E3 template [19].

The biopharmaceutical industry does not have a uniform position regarding what results need to be disclosed. Some company Web sites contain only the citation of a publication, some include the three-page synopsis from the clinical study report, others include the actual clinical study report, and others include a summary that is intermediate to a synopsis and a full clinical study report. Regardless of the length or format, industry agrees that results of hypothesis-testing/confirmatory trials should be made publicly available.

It is important that the public understand that the intent of the results database is to present the results of individual clinical trials for transparency, not to replace patient-physician interactions or the comprehensive nature of the product label. It is important for readers of the results to understand

proprietary assets for the sponsor and anything that puts these assets at risk could influence the sponsor's ability to be innovative. Early disclosure of protocols can provide competitors with clinical trial methods, and early disclosure of clinical trial results can tip off a competitor on new indications that they are pursuing.

The intent of the results database is to provide transparency/accountability; to disclose the trial's methods and its results. It was never the intent of the results database that decisions or interpretations would be based on the results of a single trial, but there is the danger that this will occur.

Most exploratory trials are proprietary to the company conducting them, and because they are typically smaller trials, they have limited statistical power and are primarily conducted for generating hypotheses for future trials.

The data presented in a summary/synopsis may not match those presented in a manuscript. Journal reviewers and editors may request different analyses than were originally performed, and as a result, there could be a slight difference in the data reported. These different post-hoc analyses should not change the overall interpretation of the findings. Some industry Web sites contain language clarifying the fact that the results presented in the summaries may be different from those presented in a manuscript. Unfortunately, in the present environment, differences may be interpreted as an indication that the sponsor has altered the reported results.

Future directions

The industry supports collaboration among the global industry, academics, regulators, legislators, and other stakeholders, to develop uniform, consistent standards across regions. The current trend for the development of separate requirements by organization and locality is neither efficient nor effective. Some biopharmaceutical companies support the discussion of disclosing trial results rather than spending much of their efforts on the registration of trials at inception. The data presented on the various Web sites differ in format and the amount of data disclosed. As stated previously, the original intent of registries and results databases was to provide patients and providers with relevant information about clinical trials that may be enrolling patients; provide patients and providers with the results of a clinical trial once the trial is completed and the product is marketed; and provide transparency that each clinical trial initiated has a reported outcome.

Uniform standards are necessary to help meet the intent of the registries and results databases. To facilitate the sharing of data and the transparency of clinical trials, internationally accepted uniform standards for posting data at trial initiation and trial completion are needed. These standards should apply to all entities conducting trials in humans without exceptions. The standards need to meet the goal of providing patients and providers with information at trial inception, while not deterring innovation or creating an undue burden on industry.

An all-or-none approach to disclosing clinical trial data from all trials may not be the answer. Although the ICMJE has stated that the only way to ensure clinical trial results are not "file-drawered" (a concept where results of trials are hidden from public view) is to demand that specific information be disclosed at initiation, it is possible that the best way to serve the public would be to list trials being conducted at inception (without jeopardizing innovation by requiring 20 fields), and then post the results in more detail than space constraints will allow for in journal publications. Registration is the starting point and study results are the endpoint; neither is meaningful without the other.

The controversy surrounding clinical trial data disclosure requires a more balanced approach. Requiring disclosure without carefully considering the potentially damaging consequences to the sponsor could decrease the number of companies willing to continue in the high-risk drug development business, which in turn will negatively influence consumer health. Clinical trial registries and results databases can be beneficial tools in the healthcare process for patients and their providers; however, there should be a balance between the public's need for information through disclosure and industry's need for protecting confidentiality and innovation.

The discussions regarding clinical trial registries and results databases are ongoing and this issue continues to evolve. As more conversations take place between stakeholders and the biopharmaceutical industry, it is likely that changes will occur in what, where, when, and how clinical trial information is disclosed.

Acknowledgements

I wish to acknowledge the contributions of Ed Campanaro, Executive Director, Clinical Operations, Cubist Pharmaceuticals, Inc., Lexington, MA and Barbara Godlew, RN, Clinical Research Manager, Clinical Trials Registry Office, Novartis Pharmaceuticals Corporation, during the review of this chapter.

References

1   United States Food and Drug Administration. Section 113 Food and Drug Administration Modernization Act of 1997. Available at: http://lhncbc.nlm. nih.gov/clin/113.html (Accessed 8 February 2006)

2 Pharmaceutical Research and Manufacturers of America. Principles on
   conduct of clinical trials and communication of clinical trial results. Available at: www.phrma.org/publications/policy/2002-07-18.490.pdf (Accessed
   8 February 2006)
3 Pharmaceutical Research and Manufacturers of America. Updated principles for conduct of clinical trials and communication of clinical trial results. Available at: www.phrma.org/publications/quickfacts/30.06.2004.427.
   cfm (Accessed 8 February 2006)
4 Pharmaceutical Research and Manufacturers of America. Clinical
   trial registry proposal. Available at: www.phrma.org/publications/poli-
   cy/06.01.2005.1111.cfm (Accessed 8 February 2006).
5 De Angelis C, Drazen JM, Frizelle FA et al. Clinical trial registration: a
   statement from the International Committee of Medical Journal Editors.
   Available at: www.icmje.org. (Accessed: 29 January 2006).
6 National Cancer Institute. PDQ-NCI's comprehensive cancer database.
   Available at: http://www.cancer.gov/cancertopics/pdq/cancerdatabase (Accessed 8 February 2006)
7 International Federation of Pharmaceutical Manufacturers and Associations. IFPMA Improves Biomedical Data Transparency with Launch
   of First Worldwide Clinical Trials Portal www.ifpma.org/News/NewsRe-
   leaseDetail.aspx?nID=3471 (Accessed 8 February 2006)
8 Pharmaceutical Research and Manufacturers of America. Joint position
   on the disclosure of clinical trial information via clinical trial registries and
   databases. Available at: www.phrma.org/publications/policy//admin/2005-
   01-06.1113.PDF (Accessed 8 February 2006)
9 WHO Technical Council on Clinical Trial Registries Standards; www.who.
   int/ictrp/news/ictrp_sag_meeting_april2005_conclusions.pdf (Accessed 30
   January 2006)
10 International Federation of Pharmaceutical Manufacturers and Associations. Clinical trials portal. Available at: www.ifpma.org/clinicaltrials.html
   (Accessed 8 February 2006)
11 De Angelis C, Drazen JM, Frizelle FA et al. Is this clinical trial fully registered? A statement from the International Committee of Medical Journal
   Editors. Available at: www.icmje.org (Accessed: 29 January 2006.)
12 Tufts Center for Drug Development. Tuft Center for the Study of Drug
   Development pegs cost of a new prescription medicine at $ 802 million.
   Available at: http://csdd.tufts.edu/NewsEvents/RecentNews.asp?newsid=6
   (Accessed 8 February 2006)
13 United States Food and Drug Administration. Challenge and opportunity
   on the critical path to new medical products. Available at: http://www.fda.
   gov/oc/initiatives/criticalpath/whitepaper.html#f5 (Accessed 8 February
   2006)

14 Harris Interactive, Inc. New Survey Shows Public Perception of Opportunity to Participate in Clinical Trials Has Decreased Slightly From Last Year. Available at: http://www.actmagazine.com/appliedclinicaltrials/article/articleDetail.jsp?id=168222 (Accessed 9 February 2006)

15 Center for Information and Study on Clinical Research Participation. Clinical trial registry survey. Available at: http://www.ciscrp.org/programs/documents/2005registrysurvey.data.forwebsite2.pdf (Accessed 8 February 2006)

16 Fair Access to Clinical Trials Act of 2004, House of Representatives, HR 5252, 108th Congress, Second Session. Available at: http://www.govtrack.us/data/us/bills.text/108/h5252.pdf (Accessed 9 February 2006)

17 Fair Access to Clinical Trials Act of 2004, US Senate, S 2933, 108th Congress, Second Session. Available at: http://www.govtrack.us/data/us/bills.text/108/s2933.pdf (Accessed 9 February 2006)

18 Pharmaceutical Research and Manufacturers of America. PhRMA Clinical Study Results Database Proposal. Available at: www.clinicalstudyresults.org/primers/Clinical_Study_Results_Database.pdf (Accessed 8 February 2006)

19 ICH Harmonised Tripartite Guideline E3, Structure and Content of Clinical Study Reports, November 1995, www.ich.org/MediaServer.jser?@_ID=479&@_MODE=GLB (Accessed 8 February 2006)

Clinical Trial Registries: A Practical Guide for Sponsors and
Researchers of Medicinal Products, edited by MaryAnn Foote
© 2006 Birkhäuser Verlag Basel/Switzerland

Public and patient usage and expectations for clinical trial registries

Kenneth Getz

Tufts Center for the Study of Drug Development, Tufts University, USA

Introduction

For more than 10 years, clinical research professionals, policymakers, regulatory agencies, legislators and patient advocacy groups have been discussing and implementing initiatives designed to make clinical trial information more accessible, transparent, and credible [1–3]. Heightened attention and urgent debate around clinical trial and trial results registries has been prompted most recently by serious allegations that the lack of transparency and accountability, even concealment, has compromised patient safety and public health [4].

Although disclosure and transparency of clinical research information by online registries is primarily intended for use by the public, to date, this active debate has not had the benefit of input from the public and patient communities. Whereas it is true that patient advocacy groups have weighed in on the debate and they typically play an important role as feedback mechanisms on behalf of the public and patient communities, arguably, input primarily through advocacy groups is limited. Distillation and filtering of attitudes and behaviors is a frequent by-product of organized committees. In addition, advocacy groups often focus their feedback on select agendas.

As a result, the breadth and depth of public feedback is often diminished or lost.

Despite two decades of experience with registry information posted online, we do not understand behaviors and preferences of online registry users. Some of the simplest and earliest versions of clinical trial registries well predate the FDAMA 1997, Section 113 mandated initiatives [5]. These registries include government-funded offerings such as PDQ (the Physician's Data Query registry of cancer clinical trials maintained by the National Cancer Institute) and ACTIS (the AIDS Clinical Trials Information Service), as well as private sector offerings such as the CenterWatch Clinical Trial Listing Service, Acurian, and VeritasMedicine. Indeed, there is a rich history of user experiences with many of these early registries, and with clinicaltrials.gov, now more than 5 years old. At the time of this writing, public and private sector registries have not shared, nor have many of them ever assessed, the experiences and needs of their actual users.

A primary goal of this chapter is to amplify the public's voice to assist in informing the debate. This chapter presents the results of an online survey among nearly 5000 visitors to recent clinical trial and trial results registries. Given the limitations of these survey results, a secondary goal of this chapter is to encourage registry hosts, both in the public and private sectors, to share what they have learned and what they continue to learn from their users.

Ideally, public input will help ensure that clinical trial and trial results registries offer a high level of value and utility. Input from the public and patient communities will be immensely helpful in determining what content to provide; how content will be used; what functionality is most useful; how to best disseminate the information; and, ultimately, how to use this initiative to help in addressing eroding public trust in the clinical research enterprise.

Methdology to assess public and patient usage

In the absence of data provided by registry hosts on user preferences and behaviors, the Center for Information and Study on Clinical Research Participation (CISCRP), an independent non-profit group based in Boston, conducted a survey among approximately 5000 registry users. The survey was conducted online during the months of June and July 2005. Prospective participants in the study were notified by a dozen Web-based registry hosts who had offered to assist CISCRP in this effort. As part of this notification process, prospective participants were asked to respond only if they had visited an online registry within the past 3 months. Prospective participants e-mailed CISCRP and indicated their willingness to complete an online survey. A total of 10 053 interested registry users were invited by CISCRP to complete the survey. Nearly half (48%) of the total invitees completed the survey, representing an excellent response rate and sample size.

Profile of respondents

Two-thirds of registry users participating in this survey are patients and one-third are family members or friends of patients. More than 80% of respondents have completed at least a portion of their college education: 13% completed graduate work; 27% graduated from college, and 39% attended a college program. The distribution of respondents by level of education is representative of the typical Internet user.

Respondent age distribution is slightly older, and sex distribution is disproportionately female, than that of the general Internet population. A high percentage of registry users (48%) responding to the CISCRP survey are aged 45 years or older. Approximately 25% are between the ages of 36 and 45 years. In addition, most respondents (79%) are women. Both age and sex distributions are consistent with and representative of people who use the Internet to search for medical and health information.

Usage of registries

Twenty-five percent of users access clinical trial and trial results registries through major commercial search engines such as Google or Yahoo. The most commonly used search terms are: 'clinical trials'; '[Disease Name] clinical trials'; and '[Drug or Treatment Name] clinical trials'. An additional 38% of users access registries through links from other Web sites, such as those listing general or disease-specific health information. Very few users, less than 10%, report book marking (i.e., capturing a favorite Web page for future reference) a given clinical trial or trial listing registry. This behavior is consistent with descriptions of study volunteers in the literature. When initially seeking clinical trial information, study volunteers and members of their support network are often characterized as harried, impulsive, and even desperate. In addition, it appears that most users locate registry information as part of a broader search for any and all information that may offer additional insight and hope into their treatment options.

Users are visiting an average of four registries per search, and it does not appear that they discriminate between general, disease-specific, and company-specific content. Each visit is brief in duration, averaging less than 5 min, to peruse clinical trial listings and results information. It is clear that registry users are looking for very specific information, regardless of source. This will be discussed in more detail.

At this time, registry users have mixed preferences about whether they would prefer a one-stop, central registry for all trial listings and results information (53%) or many disparate registries (47%). Based on coded, open-ended explanations of their preferences, the primary reason cited by most registry users (66%) is that they do not trust a single source for quality information at this time. When the question is posed differently to assess a more long-term outlook, i.e., 'How important is it that a single registry list all trials and trial results', 95% of users believe it is 'Somewhat' or 'Very' important, and this is largely due to the convenience offered by a central source.

Assessment of information quality, scope and usefulness

An oft-cited 2002 article by Manheimer and Anderson [6] characterizes well the pervasive poor and variable quality of content found on clinical trial and trial results registries. The authors conducted a meta-analysis of numerous government and commercial registries and found widespread shortcomings (Tab. 1).

Table 1. Some widespread shortcomings of government and commercial registries [6]

- Incomplete listings of clinical trials
- Missing information within the clinical trial listings
- Misspellings of drug names, diseases, research centers, and locations
- Conflicting locations where, and dates when, trials are being initiated, executed, and completed

Registry users do not appear to be anywhere near as discriminating as Manheimer and Anderson. Most registry users positively rate the amount, quality, and ease of use of registry information with few exceptions. Twenty percent of users report that finding information is difficult and 17% of users report that registry information is difficult to understand. These largely positive ratings may in part be because most registry users have no context or basis with which to compare quality, scope and utility of the information. These results are not entirely surprising, given the large number of autonomous and incomplete clinical trial and trial results registries combined with the rushed manner by which users find and scan clinical trial information.

Users also have generally realistic, though optimistic, expectations concerning the completeness of information on a registry. The typical user expects a given registry to list 40% of all active clinical trials and 50% of all clinical trial results.

According to our survey, registry users have very basic information needs. Individuals seeking active clinical trials are primarily interested in seeing summary trial information; whether the trials are being conducted in easily accessible locations; and investigative site contact information on a registry. Investigator credentials, study requirements, glossaries of terms and compensation amounts are of less interest to the user at this stage of their search. Individuals seeking clinical trial results information are looking for summary descriptions of trial outcomes and short listings of relevant citations from peer-review journals.

Along with the preferred areas for which users want digestible information, a high percentage of registry users want to understand what they should do with the information once they have found it. Eighty-seven percent of respondents said that they would like registries to provide suggestions on what to do with information once the user finds what he or she is looking for; 84% of users said that they would like registries to teach them how to evaluate the information to determine the appropriateness and risk/benefit of a given clinical trial. This feedback from registry users captures the relative importance of providing more than just clinical trial information.

Registry users have very limited knowledge about, and experience assessing and using, clinical trial information. This finding is consistent with the results of a public opinion poll conducted by CISCRP in late 2004 showing that only 5% of the American public feels confident that they would know where to find, and how to evaluate, an appropriate clinical trial.

How registry information is being used

More than 80% of registry users report that they initially printed out (52%) and recorded by hand (29%) information that they found on various registries. Approximately 10% of users initially e-mailed registry information to a family member or friend. Only 1% of registry users report initially emailing the information to their healthcare provider. Eventually, 50% of all registry users report showing the information to a family member or friend and 25% state that they showed the information to their healthcare provider (physician or nurse).

Additionally, 28% of registry users who found a 'relevant' clinical trial contacted the research center within the first 2 weeks of finding the information. Although users may be conducting their search urgently, after a recent diagnosis of a severe or life-threatening condition, a relatively small percentage immediately contacts the research center. Again, this finding may be explained in part by the fact that most registry users neither know what to do with clinical trial information

nor how to evaluate that information. Two to eight weeks later, an additional 17% of registry users who found a relevant
clinical trial contacted the research center. The low incidence
of initial contact made with the investigative site is consistent
with those observed by CenterWatch and Acurian.

Conclusions

The results of this initial study help to inform the debate across
three primary areas: what content to provide; what medium or
distribution approach to use; and what level of compliance is
required to ensure that all information is publicly accessible
online.

With respect to what content to provide: debate among
professionals has focused intensely on a variety of issues such
as the inclusion of 'hypothesis testing' *versus* 'exploratory'
clinical trials; published and unpublished clinical trial results;
and blinded data fields to protect property and competitively
sensitive information. Public and patient communities indicate
a high interest in basic, summarized, and distilled information
supplemented by instructions and guidelines on how to use it.
Further, the public and patient communities are interested in
learning how to evaluate the quality, credibility, and the limitations, of online clinical trial information.

Regarding what medium to use to disclose clinical trial information, debate has focused on centralized *versus* autonomous registries. The public and patient communities are re-
ceptive to decentralized clinical trial and trial results registries
in the short-term. Eventually, as public and patient trust in
the credibility and comprehensiveness of registry information
increases, a one-stop information source is preferred in large
part due to the convenience that this approach offers.

It is important to note that online registries provide clinical trial information transparency and accessibility to only a subset of all people who might benefit from this information. According to a 2003 report by the Pew Internet American Life Project, more than 40% of all Americans neither use nor have access to the Internet. Presently, there is a higher level of disparity in Internet usage among African-American and Latino communities with 55% and 49% being non-Web users, respectively. To extend access to all communities, registry information providers need to consider a variety of formats including print medium and word-of-mouth. A number of grass roots support groups and community centers, for example, publish and disseminate information that originated on the Internet.

Public and patient communities have not provided input into whether to mandate adherence and to provide incentives and punishments for non-adherence. Registry users openly acknowledge that no single registry is the 'go-to' comprehensive Web location. Registry users do not trust one central source, and they are accustomed to browsing numerous Web sites to gather information. According to surveys conducted by CISCRP and Harris Interactive, public trust in the clinical research enterprise has eroded significantly. Full disclosure and compliance, by all research sponsors, of clinical research information through easily accessible and user-friendly means will play an important role in repairing public trust.

Sponsors and researchers could do much more beyond disclosing information through online registries. It is essential that sponsors provide general context and public education about the clinical research process in general, and the role it plays in advancing public health. Less than 20% of patients diagnosed with severe and life-threatening illnesses report considering clinical trials as a health care option. Ultimately, only 6% of the eligible population, patients diagnosed with debilitating and life-threatening conditions, participates in clinical trials annually.

Public outreach and pre-education would go far in helping to address progressively declining levels of patient participation in clinical trials. By doing so, sponsors and researchers would improve the ability of registries to positively affect clinical trial enrollment and retention rates. Since the mid-1980s in the United States, for example, spending on patient recruitment programs by investigative sites and research sponsors has grown by 16% annually, reaching more than $500 million in 2005. In that same period, volunteer randomization rates have declined steadily. According to reports from several hundred investigative sites and analyses of study completion records, nearly 45% of volunteers were enrolled in clinical trials in 1984. In 2003, randomization rates declined to less than 25%. Whereas, approximately 17% of enrollees dropped out of clinical trials in 1984, an average of 27% dropped out of clinical trials in 2003.

The public considers their healthcare providers to be the most trusted source for medical information, yet less than 20% of volunteers report first learning about clinical trials from their physician or nurse. In a recent survey conducted among board-certified physicians in active community practices throughout the United States, less than 50% report referring their patients into clinical trials with an average referral rate for each physician of less than one patient per year. In 2004, less than 17% of the total number of volunteers completing clinical trials came directly from physician referrals, primarily due to healthcare providers lacking adequate information and context with which to make informed decisions on behalf of their patients. Based on this recent survey among registry users, only a small percentage reports sharing information with their healthcare provider. Pre-education and outreach to the medical and health professional communities will assist in engaging this vital support mechanism for patients.

Pre-education and outreach efforts in select communities have already shown much promise in dramatically increasing the rate of prospective volunteer inquiries and in increasing physician referral rates. To be effective and successful, broad education and outreach must have the support of the entire clinical research professional community. These programs must be well coordinated and integrated into all clinical research activities. Along with online registries, a comprehensive, focused set of pre-educational programs are needed to introduce the public, prospective volunteers, health professionals, the media, and policymakers to the large amount of clinical research activity being supported and conducted within the community each year. These initiatives should include educational materials and communications to assist volunteers in framing the questions to ask of the research community to protect their safety and rights; to inform their ability to understand and evaluate clinical research information; and to assist them in making the profound decision to participate in a clinical trial.

Debate among professional, regulatory and legislative communities has primarily sought to answer the question 'What clinical trial data do we want the public and patients to have?' This chapter seeks to begin to answer a very different question: "What clinical trial data and information does the public and patient most want and how will that information be used?" Public and patient community input into this latter question is essential to the planning and implementation of enduring, effective, and widely used registries of clinical trials and clinical trial results. Done correctly, online registries will play a critical role in engaging the public, and in repairing and building confidence in the clinical research enterprise.

References

1 De Angelis C, Drazen JM, Frizelle FA et al. Clinical trial registration: a statement from the International Committee of Medical Journal Editors. Available at www.icmje.org (Accessed: 29 January 2006)

2 Hayward R, Kent D, Vijan S, Hofer TP (2005) Reporting clinical trial results to inform providers, payers and consumers. *Health Affairs* 24: 1571–1581

3 Fisher CB (2006) Clinical trial results databases: Unanswered questions. *Science* 311: 180–181

4 Office of the New York State Attorney General Eliot Spitzer. Major pharmaceutical firm concealed information. Available at http://www.oag.state.ny.us/press/2004/jun/jun2b_04.html (Accessed 17 March 2005)

5 United States Food and Drug Administration. Section 113 Food and Drug Administration Modernization Act of 1997. Available at: http://lhncbc.nlm.nih.gov/clin/113.html (Accessed 8 February 2006)

6 Manheimer E, Anderson D (2002) Survey of public information about ongoing clinical trials funded by industry: evaluation of completeness and accessibility. *Br Med J* 325: 528–531

Clinical Trial Registries: A Practical Guide for Sponsors and
Researchers of Medicinal Products, edited by MaryAnn Foote
© 2006 Birkhäuser Verlag Basel/Switzerland

Building a global culture of trial registration

Karmela Krleža-Jerić

Randomised Controlled Trials Unit, Canadian Institutes of Health Research,
Ottawa, Ontario, Canada

Introduction

As clinical trials are increasingly conducted in different regions of the world, there is a need to document, evaluate, and synthesize the global body of results. Clinical trial registration has a key role in public knowledge sharing and the unbiased evaluation of health interventions. The importance of registration has been described for decades, but recent events have prompted widespread public interest and demands for increased transparency in clinical research [1–6].

Most current ongoing trials are not based on all existing knowledge, because not all knowledge is publicly available. Some important information remains hidden in different pockets owned by different entities. Some information is disclosed only after delays, sometimes as long as a decade or more [2, 7–10].

The lack of information sharing is an obstacle to further creation of knowledge [2]. Today, research is done at such a fast pace that the lack of or delay of disclosure of ongoing trial data makes it possible that different research teams are simultaneously studying the same issue in isolation and each team is working with only a part of the available evidence. Both time and resources can be wasted in this situation. More importantly, clinical trial participants may be exposed to unnecessary risk if the therapy being studied has harmful effects that have not been disclosed. Above all, clinical research can only be justified if the knowledge gained from the research is made publicly available for the public good. Only by registering trials prospectively can researchers and the general public be sure that all trials that are undertaken are also fully reported.

Physicians have long dealt with this dilemma. The Hippocrates oath must be kept in mind when considering the ethics of clinical trials. The original oath states in part: "I will keep from them those that might harm them" [11]. The Declaration of Geneva, the modern version of the Hippocratic Oath, drawn by the World Medical Association in 1948 states in part: "The health of my patient will be my first consideration" [12]. Furthermore, the Declaration of Helsinki specifically states: "The design of all studies should be publicly available" [13].

Considering that current trials tend to be multicenter and multinational studies, it is necessary to develop worldwide standards of trial registration. If we fail to do so, we shall simply helplessly watch (or discover years too late), the shifting of trials to less protected areas of the globe. Therefore, the roles of the Ottawa Group, The World Health Organization (WHO), and the International Committee of Medical Journal Editors (ICMJE) are extremely important to ensure that the same clinical trial standards apply no matter where the trial takes place.

This chapter presents the perspectives of the Canadian Institutes of Health Research (CIHR) and the Ottawa Group, and their interaction with the WHO process for clinical trial registration.

Perspectives from the Canadian Institutes of Health Research

CIHR is the federal funding agency for health research in Canada [14]. The mandate of CIHR is „to excel, according to internationally accepted standards of scientific excellence, in the creation of new knowledge and its translation into improved health for Canadians, more effective health services and products and a strengthened Canadian health care system." [14]. CIHR is not simply a funding agency that contributes to global health through research and training. It also strives to promote a high level of ethics and accountability among scientists.

During past 5 years, CIHR has supported approximately 30 new randomized controlled trials a year, with approximately 100 trials active at a time. Trials are becoming larger, multi-center, and multinational compared with trials in earlier years. More than 50% of the trials are nondrug trials.

CIHR has paid particular attention to randomized trials, forming the Randomized Controlled Trial (RCT) Peer Review Committee, a unique methodology-related peer review committee at CIHR. The RCT Unit was created at the same time.

Acting on the principal that is necessary, for scientific and ethical reasons, to know the existence and results of all clinical trials, CIHR has become a committed partner in the international drive to enhance transparency and accountability in clinical research. Three primary areas have been targeted to facilitate prospective registration of clinical trials (Tab. 1).

Table 1. *Three primary areas have been targeted by CIHR to facilitate prospective registration of clinical trials*

- Internal oversight of randomized trials funded by CIHR
- Public registration of CIHR-funded trials
- Contribution to international dialogue on trial registration

Internal oversight: The need for increased accountability, quality control, and transparency in CIHR-funded trials

In the context of the primary mandate of CIHR, its RCT Unit recognized the need to evaluate the quality and impact of CIHR-funded trials. In 2003, a comprehensive analysis was initiated to review the reporting of results of all 105 randomized trials funded from 1990 to 1998 [15]. Among these large, rigorously peer-reviewed studies, only 55% of completed trials were published and 26% were being prepared for publication at the time of the analysis. When the published journal articles based on the trials were compared with the clinical trial protocols, an average of 33% of efficacy outcomes and 60% of safety outcomes per trial were inadequately reported or omitted from the journal article. Results that were fully reported were substantially more likely to be those that showed a statistically significant difference between study groups. Furthermore, 40% of trials had different primary outcomes in the protocol compared with the primary outcomes reported in the published journal article.

Although consistent with other studies [16], these findings were the first to demonstrate suboptimal reporting of results in rigorously designed, government-funded research and confirmed a need for CIHR to establish ongoing mechanisms to monitor its funded trials through an internal trial database and public registration (Tab. 2).

Table 2. CIHR efforts to increase the quality, accountability, transparency, and impact of randomized clinical trials (RCT) from 2002 to 2006

Internal efforts
- Special RCT Peer Review Committee
- Creation of the RCT Unit
- Outcome reporting study
- The RCT database (RCT Summary Form)

Introduction of the final report – CONSORT and Summary form
Requiring registration of CIHR-funded trials
International initiatives
- Participation and funding for the Ottawa Group
- Participation and funding for the WHO International Clinical Trials
  Registry Platform

Randomized controlled trial database

The RCT Database of CIHR was established to monitor compliance with the original approved protocol and to identify the selective reporting of results. It also serves as the basis for trial registration and dissemination of trial results. The RCT Database captures the main protocol information at three distinct points in the life a trial: when submitted for approval for funding by CIHR; at interim stage(s), such as oversight or funding renewal; and at the end of trial.

Data are collected from principal investigators using a standardized summary form. Once verified by CIHR, the data are entered into RCT Database. At the interim stage, the same form is used again to indicate any protocol changes. At the end of a trial, a final report based on the CONSORT Statement [17], with the updated summary form, is submitted to CIHR. If a given trial remains unpublished, CIHR will collaborate with the principal investigator to ensure that the CONSORT-based final report will be publicly available on the Internet. Throughout this process, CIHR and the principal investigator maintain a close relationship.

From the CIHR perspective, the RCT database, oversight, and trial registration are all strongly connected. The same summary data recorded in the CIHR RCT Database are used for trial registration, oversight, and control of biases arising from selective reporting of results. Principal investigators submit their trial data only once to the funder and the trial register. The one-step trial data submission not only saves time for the investigator but also ensures that data in the CIHR database and the international register are consistent and accurate. The collaboration between CIHR and investigators in data preparation also ensures that the study does not appear multiple times in multiple trial registers. It is expected that the process will improve the quality of protocols by indicating the key methodological components of a trial (Fig. 1).

Registration of CIHR-funded trials in an international register

In late 2003, CIHR chose to register its funded trials with the International Standard Randomised Controlled Trial Number (ISRCTN) [17, 18] register because at the time it was the only international register that met all the requirements of the CIHR: it accepted all types of randomized trials, assigned a unique number, and had free access on its publicly accessible Web site [14]. ISRCTN also allowed CIHR to directly register the trials it supports and, thus, increase the possibility that accurate and meaningful data are shared. The cost associated with trial registration is paid by CIHR.

Through these actions, CIHR has assumed a unique role among public funders, going a step beyond merely demanding trial registration and struggling to ensure compliance. The experiences of others, especially the United States Federal Drug Administration (FDA), were very informative and contributed to this decision [19, 20].

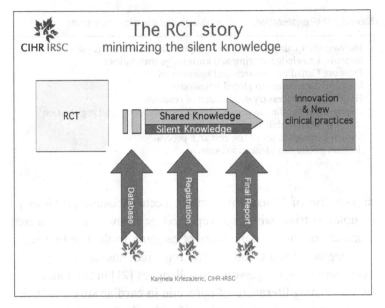

Fig. 1. The randomized clinical trial story at CIHR.

CIHR decided to use an existing international register, and to invest its time and resources in the development of international trial registration standards, rather than developing and maintaining its own national trial register. CIHR hopes to inspire the Canadian research community to register all trials conducted in Canada, and thus help create the culture of trial registration in Canada and beyond.

CIHR had specific expectations concerning the registration of trials it supports (Tab. 3). Some of the expectations are specific to CIHR and some are more general. To address the very specific one, i.e., to help avoid unnecessary duplication of research, it should be noted that this expectation is not just important to CIHR, but also to trial participants and investigators. CIHR, and probably other public funding agencies and the scientific community, is concerned with the unnecessary

Table 3. CIHR expectations from registration of the trials it supports

- Increase the quality of trials and the publication of their results
- Improve knowledge sharing and knowledge translation
- Promote Canadian research and researchers
- Identify contributors to global knowledge
- Help avoid unnecessary duplication of research
- Inspire other funding agencies to require mandatory trial registration
- Increase accountability
- Improve transparency of the research process
- Regain public confidence in research

duplication of trials, which might occur because previously completed trials were not reported or known, and because researchers often are unaware of ongoing trials. The first reason, neglected trials that are not reported, has been very well documented by Fergusson and colleagues [21] in their analysis of the existing literature of aprotinin in cardiac surgery. Their study exposed a large number of redundant trials that were addressing questions that had been definitively answered. A combination of trial registration and systemic reviews of trials would help public and other funders to optimize their resource allocation.

Table 4. CIHR has a clear vision regarding clinical trial registration

- Register all trials, starting with randomized controlled trials
- Register the minimal data set defined by WHO and more
- Contribute to the international trial registration and knowledge-sharing dialogue
- Enable electronic grant application and electronic protocol and results registration

CIHR vision regarding trial registration

CIHR has a clear vision regarding clinical trial registration (Tab. 4). Following its vision, CIHR participates actively both in the Ottawa Group international dialogue and in the WHO-led project on trial registration.

The Ottawa Group and Ottawa Statement

Inspired by its own intention to increase transparency in clinical research and the announcement that the ICMJE would only publish trials that had been registered at inception [4, 5], CIHR initiated an international meeting of interested stakeholders about trial registration during the 2004 Cochrane Colloquium. A grass-roots international dialogue was started, and has expanded beyond its Canadian origins.

The Ottawa Group consists of individuals and organizations that endorse a set of fundamental guiding principles for trial registration – the Ottawa Statement (Part 1). The international group is composed of consumers, researchers, journal editors, and representatives from ethics committees and trial registers from five continents. Members are committed stakeholders who are pushing the boundaries of knowledge disclosure in clinical research by proposing a high level of transparency. Consistent with its vision of openness and transparency, the Ottawa Group welcomes all interested individuals to participate in discussion at face-to-face meetings and at the Ottawa Group Web site [22]. It is important to emphasize that the Ottawa Group is not a Canadian, CIHR or WHO working group (Tab. 5).

Table 5. The activities of the Ottawa Group

- Promote dialogue on trial registration through face-to-face meetings, the Internet, and Web sites
- Develop principles for trial registration and its implementation
- Contribute to and support the WHO Registry Platform

As of April 2006, the Ottawa Group had produced the Ottawa Statement Part 1 on principles of trial registration [6] and a draft of the Ottawa Statement Part 2 on principles of implementation of trial registration, and has been maintaining its website, hosted by the Ottawa Health Research Institute (OHRI). Future plans involve developing further recommendations for the registration and disclosure of results.

The Ottawa Statement Part 1

The Ottawa Statement Part 1, Principles for International Registration of Protocol Information and Results from Human Trials of Health-related Interventions was published in the British Medical Journal and posted on the Ottawa group Web site. Ottawa Statement Part 1 principles are simple and clear: disclose protocol details up front, amendments as they occur and results at the end, as illustrated in (Fig. 2). They can be summarized in five major points:

1. There is a strong ethical and scientific rationale for registering all human trials involving health-related interventions, regardless of study design or intervention type. Of greatest importance is the ethical obligation to all trial participants, whose voluntary contributions and assumption of potential risk must be balanced by public dissemination of the scientific knowledge gained from the research. This obligation exists regardless of the type of trial, and arguably is greatest in early trials where the risk of potential harm from new interventions is least known.

2. All trials should have a unique identification number to enable their tracking and global identification, which will enable all study documentation, presentations, publications, and unpublished results to be linked to a single trial.

3. At a minimum, sufficient protocol details should be publicly registered and disclosed before enrolling the first par-

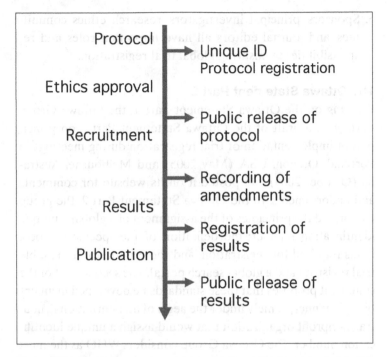

Figure 2. Ottawa Statement: general time-line for trial registration process [6].

ticipant to enable critical appraisal of the methodology and results of all trials at their inception. Amendments are registered as they occur.

4. All results for pre-specified outcomes and analyses as well as data on harms, should be publicly registered and disclosed, allowing sufficient time for publication. This disclosure will help to address the problems associated with selective reporting of results, and will ensure that the knowledge gained from research conducted on humans will serve society as a whole rather than commercial interest alone.

5. Sponsors, principal investigators, research ethics committees, and journal editors all have important roles and responsibilities in ensuring global trial registration.

The Ottawa Statement Part 2

Building on the Ottawa Statement Part 1, the Ottawa Group
developed a draft of the Ottawa Statement Part 2 on principles of implementation of trial registration during meetings in
Portland, Oregon, USA (May 2005) and Melbourne, Australia (October 2005) and posted it on its website for comments
and endorsements. In the Ottawa Statement Part 2, the group
elaborated the principles of the assignment of a globally unique
identification number; the definition of the specific protocol
items required for registration; and the criteria for acceptable
trial registries and a global search portal. The second part of the
statement proposes that these standards be developed in an unbiased manner, namely under the aegis of an international, neutral, nonprofit organization that would assign a unique identification number. The Ottawa Group considers WHO as the most
suitable body to guide and implement the process globally.

Dialogue is particularly intense concerning the trial protocol items that need to be registered and disclosed before
participant enrolment. The Ottawa Group supports the WHO
20-item Trial Registration Data Set as a good start, but finds
it insufficient to enable critical appraisal of trial methodology.
As such, the Ottawa Group proposes the expansion of that
list, particularly details of methodology and ethics approval
(Tab. 6). It is expected that this 20-plus list of items would better meet ethical and scientific needs.

Although it might seem that the Ottawa Statement Part 2
is far reaching, these items are hardly sufficient to evaluate the
methodology of trial in the process of a systematic review.

Table 6. Proposed Ottawa Group protocol items required in a clinical trial registry [22]

- Unique ID
- Secondary ID(s)
- Funding source(s)
- Primary sponsor
- Secondary sponsor(s)
- Coordinating/principal investigator
- Responsible contact person
- Official scientific title
- Simplified title for general public
- Acronym, if any
- Trial Web site
- Short simplified description (text)
- Key dates
- Ethics approval
    - Name of ethics board (REB/IRB) for the primary site in each country
    - REB trial approval number
    - Date of issue/approval
- Recruitment center locations
- Recruitment status
- Eligibility criteria
- Controlled (yes/no): If yes
    - Study design: parallel group, crossover, cluster, factorial
    - Number of arms
    - Randomized or not
        - If randomized, generation of the allocation sequence
        - If randomized, allocation concealment
    - Masking/blinding, if yes, who is blinded
    - Other design features
- Framework (superiority, noninferiority, equivalence; to be further elaborated)
- Trial objectives
- Disease/condition
- Interventions by study groups and duration
- Target sample size
- Primary outcome(s) and time point of measurement
- Secondary/additional outcomes and time point of measurement for each (including subgroup analyzes and adverse events)
- Trial phase (phase 1, 2, 3, or 4) if relevant

The impact of the Ottawa Group and the Ottawa Statement

The Ottawa Group provides a vision of the potential future paradigm of clinical research by helping to define the road ahead for improved transparency. The Ottawa Statement has triggered much discussion, from its initial and subsequent publications, particularly the comparative analysis of industry, WHO, and Ottawa Group views [2].

In the short time of its existence (since October 2004), the Ottawa Statement Part 1 has had a global impact. A list of endorsements from across the globe is continuously growing and the statement has been translated in several languages (Japanese, French, Chinese, and Spanish) [22, 23]. Furthermore, the Ottawa Statement has been used in discussions for the WHO-led trial registration process.

However, although representatives from the pharmaceutical industry have participated in Ottawa Group meetings, no pharmaceutical group or company has endorsed the Ottawa Statement. This situation is unfortunate because their endorsement would show that the industry accepts high ethical and scientific requirements of clinical research. This nonendorsement has not been expected because these principles based upon ethical, scientific, and medical grounds are difficult to challenge. Furthermore, they do not imply immediate application, i.e., endorsing them does not mean that any company would need to begin registering all details of all trials right away [2].

The WHO International Clinical Trials Registry Platform

In 2004, the WHO decided to become involved in trial registration [24]. The project was prepared by the international group of experts in a series of consultations, organized and co-ordinated by WHO, recognized at the Ministerial Summit on Health Research in Mexico in November 2004 and approved by the World Health Assembly in 2005. Trial registration was a part of the WHO commitment to implement the Mexico summit resolutions on health research. The project started in January 2005 with the aim of developing trial registration standards, enhancing access to registers, providing an advocate for compliance, and building capacity where needed [25].

After the release of the World Health Assembly resolution A58/34 3(2), WHO officially launched the project, established the project secretariat at WHO Headquarters in Geneva, Switzerland, and formed the International Advisory Board and the Scientific Advisory Group consisting of 14 and 22 members, respectively (as of April 2006) [26].

The WHO International Clinical Trials Registry Platform (ICTRP) is a global project with the aim to facilitate access to information about controlled trials and their results, including the worldwide standards of trial registration. The primary objectives of this WHO initiative are "to ensure that all clinical trials are registered and thus publicly declared and identifiable, so as to ensure that for all trials, a minimum set of results will be reported and made publicly available" [26]. With a help of its Scientific Advisory Group (SAG), the International Advisory Boards (IAB), and numerous broad consultations, including the open comments sessions over the Internet, WHO has been working on defining international trial registration standards. The work includes defining the trials and information that should be made publicly available at the onset of trials; establishing trial unique identification number and a search portal; and determining the characteristics of registries, registration, and disclosure of results. At this time (April 2006), WHO is concentrating on the issue of clinical trial registration; the issue of disclosure of results will be dealt with later.

WHO trial registration aims at registering all interventional trials. An interventional trial is "any research study that prospectively assigns human participants or groups of humans to one or more health-related interventions to evaluate the effects on outcomes. Interventions include but are not restricted to drugs, cells and other biological products, surgical procedures, radiological procedures, devices, behavioral treatments, process-of-care changes, preventive care, etc." [26]. At the stakeholders meeting in April 2005, WHO elaborated the minimal data set, consisting of 20 items, that are to be registered prospectively, at the time of the assignment of the unique identifier (Tab. 7). These 20 items complement each other and arguably provide the very minimum information needed to understand an ongoing trial. This list was consecutively adopted by the ICMJE [27, 28].

However, the pharmaceutical industry challenged this list and demanded to be able to exclude items 10, 13, 17, 19, and 20 from upfront disclosure. It is obvious that when these items are excluded the whole registration does not make much sense as illustrated by cancer trial example of a breast cancer trial (Tab. 8) [2]. Namely, as 1 of these items is the unique ID and 9 are administrative, only 10 items are trial descriptors. If the disputed 5 items are not disclosed, only 5 trial descriptors remain: brief (lay) title, condition studied, key inclusion criteria, study type, recruitment status. They alone are not sufficient to provide meaningful information about any given trial and would reduce the trial registry to a mere recruiting tool and would make trial registration meaningless. If the complete 20 items trial data are not publicly available, although provided to the register, it will be impossible for anyone but a few privileged (register, editors) to understand the trial.

Table 7. Minimum data set requirements for registration of clinical trials according to the WHO. Five items (10, 13, 17, 19, and 20) are in bold because of the initial controversy surrounding their inclusion [26].

1. Unique trial number
2. Registration date
3. Secondary IDs
4. Funding source(s)
5. Primary sponsor
6. Secondary sponsor
7. Responsible contact person
8. Research contact person
9. Brief title of study
10. **Official scientific title**
11. Countries of recruitment
12. Condition
13. **Intervention(s)/comparator**
14. Key inclusion/exclusion criteria
15. Study type
16. Start date
17. **Target sample size**
18. Recruitment status
19. **Primary outcome**
20. **Key secondary outcome**

Table 8. An imaginary breast cancer study with only the remaining trial descriptors [2]

WHO item no.	Remaining trial descriptors	Potential trial description
1	Unique trial number	12345
9	Brief trial title	New Breast Cancer Therapy
12	Condition	Breast cancer
14	Key inclusion/exclusion criteria	Women aged 35–65 years, nonsmokers
15	Study type	Double-blind, parallel placebo
16	Anticipated start date	July 2005
18	Recruitment status	Currently recruiting

Following a thorough international, Internet based open discussion which culminated with the formal face-to-face consultation in April 2006, in Geneva, the WHO announced 19th of May that new standards of registration of trials on human beings comprise the full upfront disclosure of all defined 20 items [26] for all intervention trials. This certainly opens a new chapter in clinical trials conduct.

The WHO claims that it has no intention in forming a new global registry, but wishes to build on the network of existing ones that would meet carefully defined criteria. The WHO does intend, however, to provide the unique search portal and work on defining the process of removing duplicate listing – through a deduplication process, providing unique identifiers, and providing a unique search portal.

The WHO also intends to organize a global network of primary registries, whose members will meet certain well-defined criteria, mainly being able to register all 20 items as the minimum data set, as well as updates, perform verification of quality of submitted entries, perform internal deduplication, and submit data to the WHO. These primary registries will receive trial data information directly from associate registries. However, as trials may be registered in more than one primary register, an inter-register, global deduplication exercise will be performed. After the exercise, the WHO will assign each trial its Universal Trial Reference Number (UTRN). Each single, unique trial will thus have one UTRN, and each UTRN will define a single, unique trial worldwide [26]. The number of primary registeries is not yet defined, but it is expected that a relatively small number of international, national, or regional registries will form the network of primary registries. The WHO register network will also consist of associate registers. They will be either broadly based or restricted in scope (e.g., a specific disease, company, or institution), and they will forward their data to a designated primary register for further processing.

Primary registries will provide the WHO-defined minimum data set in English, while associate registries might function in any language.

Interactions between CIHR, Ottawa Group, and the WHO trial registration project

Although CIHR, Ottawa Group, and the WHO trial registration project are very different and function independently, interactions between them have been intense and fruitful. All three are interested in improving the health of people by increasing the knowledge creation and knowledge sharing, and they identified trial registration as a powerful tool to contribute to this goal. Reflective of broader trends in global research, trials are getting larger and include multiple centers in multiple countries. The idea of a global need to deal with trial registration is obviously shared by CIHR, Ottawa Group and the WHO. The trial-related positions of the three groups are compared in Table 9.

The Ottawa Group conducts international grass roots, consensus-building dialogue, while the WHO leads the intergovernmental process. Messages coming from the Ottawa Group dialogue have been contributing to the WHO trial registration project from its very beginnings, including meetings in New York (October 2004), Mexico (November 2004), and Geneva (April 2005).

The Ottawa Group commends and supports the efforts of the WHO, but, unlike the ICMJE, it considers the minimum data set a good start but insufficient, and proposes its further expansion.

Table 9. Comparison of CIHR, Ottawa Group and WHO functions in trial registration process

CIHR	Ottawa Group	WHO
Opted to contribute to the international trial registration dialogue, rather then to create the Canadian registry	Is developing statements on principles of trial registration and their implementation as well as of registration and disclosure of results	Sets international norms and standards for trial registration and reporting that uphold scientific and ethical principles
Participates in both Ottawa Group and WHO process	Worldwide grass-root dialogue of stakeholders; contributes to the WHO process	Consensus-based process of often diametrically different interest groups

CIHR, as the Canadian federal government-funding agency of health research, registers trials, contributes to the international trial registration and has been inspiring others to embrace the culture of trial registration. CIHR acknowledges the importance of the global knowledge-sharing process linked to the growing international, worldwide conduct of research.

As CIHR made a strategic decision to invest in the international registration dialogue, it has been an active participant in the WHO-led global process and it is involved in the Ottawa Group. CIHR considers both WHO and Ottawa Group dialogues powerful international processes to contribute to and learn from. Furthermore, CIHR considers the WHO as the most appropriate international body to lead a comprehensive and global development of the international trial registration system. Such an achievement would have a major impact on clinical research, taking it to the next level. Therefore, CIHR is highly engaged in the WHO process, by supporting it and actively participating in its work, in both the International Advisory Board and the Scientific Advisory Group.

Conclusion

In this chapter, we tried to present the ongoing evolution to a new paradigm of health research, where trial registration is an integral element. We propose to see the trial registration culture as a step towards new responsible ways of conducting trials and sharing their results. The worldwide process of trials registration involves numerous stakeholders whose perspectives are presented in other chapters, while we concentrated on CIHR, Ottawa Group and the WHO.

Although perceptions around issues of trial registration are changing exceptionally fast, the changes necessarily take time to be defined and implemented. It is clear that all parties will need time to adapt and learn to function successfully in the context of the trial registration culture.

Trial registration is a tool that will improve research transparency. Its goal is to fulfill ethical and scientific obligations by disseminating knowledge gained from research, thus providing stakeholders with complete information about healthcare interventions and helping to avoid unnecessary duplication of research. Its goal is not to endanger intellectual and proprietary rights.

The benefits that trial registration will bring to society are numerous; some of them will be specific for each stakeholder's group, some shared by several of them.

At this point, the main group of stakeholders challenging the WHO proposed implementation is the pharmaceutical industry. However, they too will benefit from clinical trial registration. These benefits will include regaining public confidence, shifting their resources into more new areas, and maximizing their resources.

Society has been demanding more responsible and transparent behavior in clinical research. Public funding agencies such as CIHR can continue to have an important role in leading international dialogue in this important process. The Ottawa Group, as a grass-root group of various stakeholders, has an important role in exploring the boundaries, while the WHO is the logical, international organization that can make it happen worldwide. Thus, there is a role for each stakeholder and a benefit that will follow.

Acknowledgements
I would like to thank Mark Bisby, Vice President for Research, as well as An-Wen Chan and Isabelle Schmid, Randomized Controlled Trials Unit, Canadian Institutes of Health Research, CIHR, Ottawa, for their useful comments on several drafts of this chapter.

References

1   Office of the New York State Attorney General Eliot Spitzer. Major pharmaceutical firm concealed information. Available at http://www.oag.state.ny.us/press/2004/jun/jun2b_04.html (Accessed 17 March 2005)

2   Krleža-Jerić K (2005) Clinical trial registration: the differing views of industry, the WHO, and the Ottawa Group. *PLoS Med* 2: e378.

3   Goodyear M (2006) Learning from the TGN 1412 trial. *Br Med J* 332: 677–678

4   Abbasi K (2004) Compulsory registration of clinical trials. *Br Med J* 329: 637-638

5   DeAngelis CD, Drazen JM, Frizelle FA et al (2004) Clinical trial registration: a statement from the International Committee on Medical Journal Editors. *JAMA* 292: 1363–1364

6   Krleža-Jerić K, Chan AW, Dickersin K, Sim I, Grimshaw J, Gluud C (2005) Principles for international registration of protocol information and results from human trials of health related interventions: Ottawa statement (part 1). *Br Med J* 330: 956–958

7   Dong BJ, Young R, Rapaport B (1986) The nonequivalence of thyroid products. *Drug Intell Clin Pharmacol* 20: 77–78

8   Eckert CH (1997) Bioequivalence of levothyroxine preparations: industry sponsorship and academic freedom. *JAMA* 277: 1200–1201

9 Rennie D (1997) Thyroid storm. *JAMA* 277: 1238–1243

10 Spigelman MK (1997) Bioequivalence of levothyroxine preparations for treatment of hypothyroidism. *JAMA* 277: 1199–1200

11 Hippocratic Oath. Available at http://www.hal-pc.org/~ollie/hippocratic.oath.html (Accessed 16 April 2006)

12 Declaration of Geneva. http://www.donoharm.org.uk/gendecl.html (Accessed 17 April 2006)

13 Declaration of Helsinki. Ethical principles for medical research involving human subjects. Available at http://www.wma.net/e/policy/pdf.17c.pdf (Accessed 31 March 2006)

14 CIHR Web site. Available at http://www.cihr-irsc.gc.ca (Accessed 31 March 2006)

15 Chan A-W, Krleža-Jerić K, Schmid I, Altman DG (2004) Outcome reporting bias in randomized trials funded by the Canadian Institutes of Health Research. *CMAJ* 171: 735–740

16 Chan A-W, Hróbjartsson A, Haahr MT, Gøtzsche PC, Altman DG (2004) Empirical evidence for selective reporting of outcomes in clinical trials. *JAMA* 291: 2457–2365

17 Moher D, Bernstein A (2004) Registering CIHR-funded randomized controlled trials: a global public good. *CMAJ* 171: 750–775

18 International Standard Randomised Clinical Trial Number. Available at http://isrctn.org (Accessed 15 April 2006)

19 Derbis J, Toigo T, Woods J, Evelyn B, Banks D (2003) FDAMA 113:Information program on clinical trials for serious or life threatening diseases: update on implementation. Ninth Annual FDA Science Forum; April 24, 2003, Washington, DC. Available at http://www.fda.gov/oashi/clinicaltrials/section 113/biolaw.html (Accessed 30 March 2006)

20 Toigo T (2004) Food and Drug Modernization Act (FDAMA) Section 113: Status report on implementation. *J Biolaw Bus* 7. From: http://www.fda.gov (Accessed April 2006)

21 Fergusson D, Glass KC, Hutton B, Shapiro S (2005) Randomized controlled trials of aprotinin in cardiac surgery: Could clinical equipoise have stopped the bleeding? *Clin Trials* 2: 218–229

22 Ottawa Group Web site. Available at http://ottawagroup.ohri.ca (Accessed 1 April 2006)

23 Krleža-Jerić K, Chan AW, Dickersin K, Sim I, Grimshaw J, Gluud C (2005) The Ottawa Statement, Part One: Principles for international registration of protocol information and results from human trials of health-related interventions. *Jpn Pharmacol Ther* 33: 544–548

24 Evans T, Gülmezoglu M, Pang T (2004) Registering clinical trials: an essential role for WHO. *Lancet* 363: 1413–1414

25 Gülmezoglu AM, Pang T, Horton R, Dickersin K (2005) WHO facilitates international collaboration in setting standards for clinical registration. *Lancet* 365: 1829–1831

26 World Health Organisation Web site. Available at http://www.who.int (Accessed 31 March 2006)

27 Abbasi K, Godlee F (2005) Next steps in trial registration; minimum criteria have been agreed, and intentions restated. *Br Med J* 330: 1222-1223

28 De Angelis C, Drazen JM, Frizelle FA et al (2005) Is this clinical trial fully registered? – A statement from the International Committee of Medical Journal Editors. *N Engl J Med* 325: 2436–2438

The Japanese perspective on registries and a review of clinical trial process in Japan

Hisako Matsuba[1], Takahiro Kiuchi[2], Kiichiro Tsutani[3], Eiji Uchida[4] and Yasuo Ohashi[5]

[1]EPS Co., Ltd., Shinjuku-ku, Tokyo, Japan; [2]University Hospital Medical Information Network, The University of Tokyo, Tokyo, Japan; [3]Graduate School of Pharmaceutical Sciences, The University of Tokyo, Tokyo, Japan; [4]Second Department of Pharmacology, School of Medicine, Showa University, Tokyo, Japan; [5]School of Health Sciences and Nursing, The University of Tokyo, Tokyo, Japan

Introduction

The University Hospital Medical Information Network (UMIN) has been operating as Japan's first clinical trial registry [1], known as UMIN-CTR, since June 2005 [2]. UMIN is a public information network for medical professionals, established within the national university system. Besides operating the clinical trial registry, UMIN has been greatly involved in academic clinical research in Japan by establishing the Internet Data and Information Center for Medical Research (IN-DICE), which acts as a central registration and allocation site for clinical trial participants.

Regulations for clinical trials in Japan

After World War II, the concept of Good Clinical Practice (GCP) was developed in the United States of America and

Europe, in remorse for past human experiments and drug disasters, to ensure the ethical and scientific appropriateness of clinical trials. The concept of GCP was legislated to regulate the conduct of clinical trials. In the United States and Europe, a test article used in a study in which a novel substance or a new indication is evaluated in humans is defined as an Investigational New Drug (IND) (United States) or Investigational Medicinal Product (IMP) (Europe), and the study is an investigation. Regardless of whether a New Drug Application (NDA) will be submitted to a regulatory agency or not, trials conducted in humans are primarily regulated by legislation. Japan, on the other hand, does not have regulations that comprehensively regulate trials conducted in humans and that serve as GCP. Legislation for the regulation of clinical trials to be submitted in an NDA was first established in Japan in line with the activities of the International Conference on Harmonization (ICH). Only such trials are strictly regulated by legislation [3, 4], compliant with the so-called ICH-GCP [5], established according to the ICH agreement. Ethical guidelines for clinical trials that will not be submitted in an NDA [6-8] were issued in 2002 and 2003. These guidelines, however, are not legislation and they lack the strictness of legislation, although the concept of GCP was incorporated into the guidelines. Regulations and guidelines for clinical trials to be submitted in an NDA and other types of clinical trials are discussed in subsequent sections of the chapter. Major regulations and guidelines in Japan lag behind the United States and Europe (Tab. 1).

Clinical trials for NDA submission
The basic legislation that regulates clinical trials to be submitted in an NDA in Japan is included in the Pharmaceutical Affairs Law [9], Standards for the Conduct of Clinical Trials

Table 1. History of establishing regulations and guidelines for clinical trials in Japan, U.S, and Europe

	Global movement	Europe (UK and EC/EU)	U.S.	Japan
1875		(UK) Food and Drugs Act		
1938			Federal Food, Drug and Cosmetic Act	
1943				Old Pharmaceutical Affairs Law
1947	Nuremberg code			
1960				Pharmaceutical Affairs Law (Law No. 145)
1961	Thalidomide case			
1962			Revision of Federal Food, Drug and Cosmetic Act Kefauver-Harris Amendment Act	
1963				Standard for new drug approval tightened
1964	Declaration of Helsinki			
1965		(EC) Council Directive 65/65/EEC on the approximation of provision laid down by Law, Regulation or Administrative Action relating to proprietary medicinal products		
1966			Public Health Service: Policy on Protection of Humans Subjects	

Table 1. (continued)

	Global movement	Europe (UK and EC/EU)	U.S.	Japan
1967		(UK) Royal College of Physician: Guidelines on the practice of ethics committees in medical research involving human subjects		"Outline of Manufacturing Approval for Pharmaceutical Products" notification
1968		(UK) Medicines Act: Requiring Clinical Trial Certificate for conduct of clinical trial		
1974			National Research Act	
1975	Declaration of Helsinki revised	(EC) Council Directive 75/318/EEC on the approximation of the laws of Member States relating to analytical, pharmaco-toxicological and clinical standards and protocols in respect of the testing proprietary medicinal products Regulation or Administrative Action relating to proprietary medicinal products		
1979			Ethical Principles and Guidelines for the Protection of Human Subjects of Research (Belmont Report)	Pharmaceutical Affairs Law revised: legislation on clinical trials for NDA and notification of its plan

Year				
1989				"Standards for the Conduct of Clinical Trials on Drugs" notification: Old GCP for pharmaceutical products
1990	- International Conference on Harmonization (ICH) established - Synthroid scandal	(EC) Committee for Proprietary Medicinal Products (CPMP) Good Clinical Practice for trials on medical products in the European Community		Old GCP for pharmaceutical products enforced
1991		(EC) Commission Directive 91/507/EEC modifying the Annex to Council Directive 75/318/EEC (UK) Department of Health: "Local Research Ethics Committees" (Health Service Guidelines (91) 5)	Code of Federal Regulation Title 45: Public Welfare, Part 46 Protection on Human Subject	
1992				"Standards for the Conduct of Clinical Trials on Medical Devices" notification: Old GCP for medical devices
1993	- Cochrane Collaboration founded - Council for International Organizations of Medical Sciences (CIOMS) International Ethical Guideline for Biomedical Research Involving Human Subjects			

Table 1. (continued)

	Global movement	Europe (UK and EC/EU)	U.S.	Japan
1996	ICH: Guideline for Good Clinical Practice			
1997	- Attempt of Medical Editors Trial Amnesty - ICH: General Considerations for clinical rials	(UK) Department of Health: "Ethics Committee Review of Multi-Center Research" (Health service Guidelines (97) 23)	FDA Modernization Act	"Standards for the Conduct of Clinical Trials on Drugs" ordinance: legislation of new GCP that complies with ICH-GCP
1998		(UK) - Medical Research Council: Guidelines for Good Clinical Practice in Clinical Trials - *meta*Register of Controlled Trials launched - NHS National Research Register launched		
2000	Declaration of Helsinki revised	(UK) The Association of the British Pharmaceutical Industry: "Pharmaceutical industry leads world in launching scheme to register clinical trials in UK"	ClinicalTrials.gov launched	
2001		(EU) Directive 2001/20/EC on the approximation of the laws. Regulations and administrative practice in the conduct of clinical trials on medicinal products for human use		

Year			
2002		PhRMA: "Principles on the Conduct of Clinical Trials and Communication of Clinical Trial Results"	- Pharmaceutical Affairs Law revised: introduction of investigator-sponsored clinical trials for NDA - "Guideline for Clinical Studies on Gene Therapies" - "Ethical Guideline for Epidemiological Studies"
2003			"Ethical Guideline for Clinical Studies"
2004	- WHO: ISRCTN registration promoted - Paroxetine scandal - ICMJE statement - Ottawa Statement	Fair Access to Clinical Trials Act submitted PhRMA: Clinical Study Results Database launched	
2005	IFPMA, EFPIA, PhRMA, JPMA: "Joint Guideline for Registration and Disclosure of Clinical Trial Information" and Clinical Study Portal Site established		- "Standards for the Conduct of Clinical Trials on Medical Devices" ordinance: legislation of new GCP that complies with ICH-GCP - UMIN-CTR launched - JapicCTI launched
2006			Japanese Medical Association's registry launched

on Drugs [3], and Standards for the Conduct of Clinical Trials on Medical Devices [4], with the last two standards serving as GCP in Japan. Adverse drug reactions that occur during clinical trials are reported according to the Enforcement Regulations of Pharmaceutical Affairs Law [10], the subordinate regulations of the Pharmaceutical Affairs Law.

The Pharmaceutical Affairs Law defines two classifications of clinical trials for an NDA: company-sponsored and investigator-sponsored, the latter added in 2002 to promote clinical trials on products or indications that are unlikely to be studied in company-sponsored trials because they are not considered marketable. Pharmaceutical Affairs Law and GCPs do not apply to investigator-sponsored clinical trials that will not be submitted in an NDA conducted by academic investigators (Fig. 1).

The Pharmaceutical Affairs Law specifies that clinical trials for an NDA must provide notice of clinical trial plans to the Minister of Health, Labour and Welfare (MHLW); that clinical trials be conducted in compliance with GCP; that the MHLW review the clinical trial plan within 30 days of its submission; and that adverse drug reactions associated with the clinical trial be reported. The GCP used in NDA clinical management is established separately for pharmaceutical products [3] and medical devices [4], in accordance with ICH-GCP.

The Japanese regulatory approval system for pharmaceutical products was tightened in 1963 after the thalidomide case to require animal experiments to confirm fetal safety and to require submission of highly objective results, e.g., data obtained in double-blind trials. The Outline of Manufacturing Approval for Pharmaceutical Products notification [11] was issued in 1967; this notification clearly specifies the scope of data to be included in the application dossier for manufacturing or importing pharmaceutical products by product classification. It requires data to be reliable, i.e., data must have been pre-

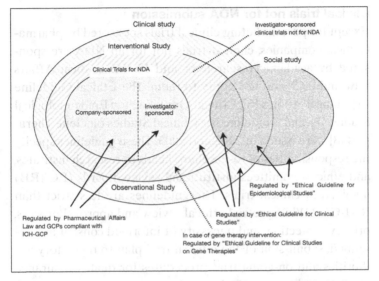

Figure 1. Clinical trial-related regulations/guidelines in Japan and their subjects.

sented at a scientific meeting or published in medical journals in Japan (the Publication in Academic Journal requirement, almost abolished in 1999 and discussed in later sections). In 1979, the legal definition of a clinical trial for an NDA (Chiken) was established. Japan's first GCP [12] was issued in 1989 and was enforced in 1990 with less strict standards compared with the current GCP [3, 4]. For example, the responsibility of sponsors to monitor or audit, or that of investigators to accept monitoring and auditing are not defined, and oral consent of clinical trial participants became acceptable. With the establishment of the present GCP for pharmaceutical products that comply with ICH-GCP in 1997 [3], standards for the conduct of clinical trials that conform to the global scientific and ethical standards were introduced in Japan. ICH-GCP-compliant standards were also established for clinical trials of medical devices in 2005 [4].

Clinical trials not for NDA submission

Except for postmarketing clinical trials sponsored by pharmaceutical companies, clinical trials not for an NDA are sponsored by academic researchers, and Pharmaceutical Affairs Law and GCPs do not apply to them. The Ethical Guideline for Clinical Studies [6], Ethical Guideline for Epidemiological Studies [7], and Guideline for Clinical Studies on Gene Therapies [8] were issued in 2002 and 2003. These guidelines specify the responsibilities of researchers, heads of research institutes, and ethics committees/institutional review boards (EC/IRB) involved in clinical trials. The guidelines are less strict than ICH-GCP. While pretrial ethical review and approval process, privacy protection, and necessity of informed consent are described, submission of the clinical trial plan to regulatory authorities and on-going trial procedures for quality assurance of trials, such as monitoring and auditing, are not discussed. The lack of strict guidelines may be because investigator-sponsored clinical trials not aimed at submitting an NDA are mainly small scale, and because of shortage of staffing and funding in Japan. Forcing full compliance with standards equivalent to ICH-GCP may be impractical at this time. Efforts were made to conduct investigator-sponsored clinical trials not aimed at submitting an NDA before the ethical guidelines were established; however, the compliance level for individual trials varied. Such varying compliance is expected to persist after issuance of the ethical guidelines.

Number of clinical trials conducted in Japan

While the Pharmaceutical Affairs Law requires submission of notice of clinical trial plans for an NDA to the MHLW, the requirement does not apply to investigator-sponsored clinical trials not involving an NDA. Tracking the number of clinical trials conducted in Japan, therefore, is difficult. The approximate number should be determined in light of the current controversy about the disclosure of clinical trial information. Estimates can be made based on historical data.

The number of clinical trials for an NDA for pharmaceutical products is known, as the number of submitted notices is published [13] (Tab. 2). According to published figures, 350–400 clinical trials are conducted every year in Japan.

Approximately 1000 investigator-sponsored, randomized, clinical trials and controlled clinical trials are conducted each year in Japan [14, 15]. This number comes from the Japan Handsearch and Electronic Search (JHES) project started in 2000 [16]. The objective of JHES is to collect the results of randomized and controlled clinical trials, including those other than original articles available in Japan, by the method described in the Cochrane Collaboration, and to develop a database to provide relevant information. Clinical trials other than randomized and controlled clinical trials, single-arm trials, and nonrandomized trials are not included in the project. Types of literature include original articles, short communications, conference proceedings, presentations made at scientific meetings, review articles, special features, serials, and roundtable discussion minutes that are published in medical journals, and various reports in Japan.

Compared with data from other countries, it is assumed that 40% of trials conducted in Japan are not found by JHES [17–19]. The number of clinical trials that do not have published results was not estimated. According to Easterbrook et al. [17], the results of approximately 50% of clinical trials are published. Considering that JHES counts presentation of results at scientific meetings and other types of literature, the percentage of clinical trials with published results may be 70%. Using this information, it is believed that 2300 investigator-sponsored trials not for an NDA are conducted in Ja-

Table 2. Number of notices of clinical trial plan on pharmaceutical products

Year	Initial notice[a]	Notice of n-th time[b]	Notice of cancellation	Notice of development cancellation[c]
1990	138	995	37	--
1991	124	1058	52	--
1992	129	1054	42	--
1993	160	1040	43	--
1994	115	965	65	--
1995	104	678	63	--
1996	95	627	64	--
1997	82	418	72	34
1998	65	341	76	67
1999	64	327	57	89
2000	80	383	41	101
2001	62	362	29	59
2002[d]	65	357	28	68
2003[d]	64	318	32	38
2004[d]	76	330	38	58

[a]Initial notice of clinical trial plan on one investigational product.
[b]Notice of newly conducted clinical trial plan on the investigational product of which initial notice has been done. "Initial notice" plus "Notice of n-th time" minus "Notice of cancellation" will be the total number of notices newly started in 1 year.
[c]Required since 1997.
[d]Compilation made between March and April in the following year.

pan yearly. With the addition of 400 clinical trials for an NDA, approximately 2700 clinical trials are conducted each year in Japan. Research is now in progress to estimate the number of clinical trials more exactly.

Publication of clinical trials in Japan

Clinical trials for NDA submission

New drug and medical device approval information in Japan is posted on the Web site of the Pharmaceuticals and Medical Devices Agency (PMDA) [20]. Evaluation reports of new drugs prepared by the MHLW and summaries of application dossiers prepared by pharmaceutical companies have been provided on the Web site since 1999. For medical devices, only evaluation reports prepared by the MHLW are posted, and the posting started in 2001. Although the information on the Web site does not include detailed results of individual clinical trials, they do summarize the results of the clinical trials.

Before approval information became public on this Web site, inclusion of reliable data, such as those provided at domestic scientific meetings or published in medical journals, in the NDA dossier had been required [11]. However, the Publication in Academic Journal requirement was almost abolished as it was not mentioned in the NDA notification issued in 1999 [21]. Four reasons for the abolition were given [22].

- Previously, around 1967, inadequate staffing had been available to ensure that the submitted data were reliable, so the peer-reviewed literature that was considered reliable was requested to be attached to the NDA; this literature became unnecessary after the domestic approval system improved to keep pace with the ICH process [21].
- It was noted that results included in the application dossier could be revised after discussion of problems noted in the evaluation process; and that the results in the literature and in the dossier could be different.
- Approval might be delayed if the applicant held the NDA until the data were published.

- The NDA might be submitted outside Japan by a competitor company using results found in the published literature, which would undermine the intellectual property of the sponsoring company.

The Publication in Academic Journal requirement was the driving force for publishing results of clinical trials on new drugs conducted in Japan [23]. Publication of clinical trial papers on new drugs in academic journals, called Chiken journals (clinical trial for an NDA journals), which served as publication media for results of this type of trial, has decreased substantially starting in the late 1990s [23]. The decrease of paper publication can be explained by the introduction of the ICH-GCP, which placed a larger burden on pharmaceutical companies to conduct appropriate clinical trials (Tab. 2), and because the Publication in Academic Journal requirement was abolished. The abolishment of the publication requirement was protested as it was a unique system and because not all results of clinical trials in other countries were published [24]. Some critics believe that the significance of the requirement should be restored given the current controversy.

The Japan Pharmaceutical Manufacturers Association (JPMA) announced a joint statement for the disclosure of clinical trial results with the International Federation of Pharmaceutical Manufacturers and Associations (IFPMA), the European Federation of Pharmaceutical Industries and Associations (EFPIA), and the Pharmaceutical Research and Manufacturers of America (PhRMA) [25]. The joint statement specifies that results of all clinical trials, other than exploratory trials, conducted on drugs approved and sold in at least one country should be publicly disclosed on a free and accessible Web site. JPMA member companies have begun to disclose summaries of their clinical trials in accordance with this joint statement.

However, trial information concerning pharmaceutical products of which development was terminated may never be disclosed. Approximately 50 notices of development cancellation are submitted annually in Japan (Tab. 2). In some cases, several clinical trials may have been conducted on a product before its cancellation. Assuming 2 clinical trials were conducted on each product, the data from approximately 100 clinical trials may not be made public.

Clinical trials not for NDA submission

The fate of clinical trials after EC/IRB approval was examined outside Japan, and suggested that many trials were not completed and the results were not published [17–19]. A study of the fate of clinical trials in Japan has not been done, but the results are probably equivalent. Assuming approximately 2300 investigator-sponsored studies not for an NDA are conducted annually in Japan and that 40% are not published, the results of approximately 900 clinical trials are unknown. JHES has researched the contents of published results [26]. Approximately 75% of clinical trials conducted in Japan were reported in Japanese, and 25% were reported in English, suggesting published results cannot be used easily in different countries because of language barrier. Japanese researchers need to consider the language to be used to publish clinical trial results to make them readily available outside Japan. To make them more available, the Database of Japanese Randomized Controlled Trial project has sent about 3300 reports of clinical trials conducted in Japan to the Cochrane Library/CENTRAL [27]. Some clinical trials using the INDICE service of the UMIN, central registration/allocation service of participants, failed to recruit enough participants. Such clinical trials are likely to be discontinued. It is well known that negative results and results considered insignificant by researchers tend not to be published. We should be aware that discontinuation of clinical trials without a clear reason also results in nondisclosure of trial results. Preparing a manuscript is difficult if the trial does not proceed as planned and is eventually discontinued. Such manuscripts are rarely published, or even submitted to a journal. Trial discontinuation needs to be made known to the public somehow. Clinical trial registries may save such trials. A data item, Major Results, is available in UMIN-CTR in anticipation of including whatever information is available from discontinued trials.

Status of clinical trial registration

Although clinical trial registration in Japan has been discussed occasionally in journals and at scientific meetings, consistent attention was not paid until the International Committee of Medical Journal Editors (ICMJE) issued a statement in 2004 [28]. Efforts were made to disclose clinical trial plans in specific areas of research using a Web site [29,30]. However, the clinical trial information provided in Web sites was not compiled into a database. Unfortunately, these sites were not good registries, as the information could not be searched electronically.

The ICMJE statement effectively alerted Japanese investigators. The statement enhanced the need for a clinical trial registration process and demanded that Japan create its own clinical trial registry. UMIN and the Japan Pharmaceutical Information Center (JAPIC) started to establish a clinical trial registries, and MHLW began to be interested in it.

Japan is behind the United States and Europe in establishing a clinical trial registry. While a number of clinical trial registries, including ClinicalTrials.gov, are available to accept registrations worldwide, Japan needs to establish its own registry for several reasons:

-   Information on clinical trials conducted in Japan should be comprehensively managed in Japan and used effectively.

Table 3. Clinical trial registries currently in operation in Japan

Registry	Organization	Main target clinical trials
JapicCTI	JAPIC	Company-sponsored clinical trials for NDA (on pharmaceutical products)
JMA Clinical Trial Registry	JMA	Investigator-sponsored clinical trials for NDA (on pharmaceutical products and medical devices)
UMIN-CTR	UMIN	Investigator-sponsored clinical trials not for an NDA

JMA: Japan Medical Association; JAPIC: Japan Pharmaceutical Information Center; UMIN: University Hospital Medical Information Network.

Information on details and status of clinical trials in Japan should be centralized and used in Japan. Domestic clinical trial information will be scattered to overseas' clinical registries if Japan does not establish its own registry.

- Information should be provided in Japanese for Japanese people for easier understanding. Registries operated in Japanese are necessary to effectively provide information to the public. Making information available to the public is one of the most important reasons for a clinical trial registration process.

- Registries in Japanese would be seen as user friendly to the Japanese researchers and would promote clinical trial registration in Japan.

Clinical Trial Registries in Japan

As of February 2006, three clinical trial registries are operating in Japan: UMIN-CTR [2], JapicCTI of JAPIC [31], and the registry of the Japan Medial Association (JMA) [32]. Each registry accepts different types of clinical trials (Tab. 3). JAPIC is a foundation with close ties to the pharmaceutical industry and operates JapicCTI primarily to accept registration of clinical trials conducted by pharmaceutical companies in line with the activities of IFPMA and JPMA. JMA promotes investigator-sponsored clinical trials for an NDA and expects to accept registration of such trials. Investigator-sponsored clinical trials not for an NDA have been registered primarily at UMIN-CTR as UMIN has been very involved with academia; UMIN-CTR will accept any type of clinical study, including clinical trials for an NDA and observational studies. While JapicCTI and JMA registries primarily accept registration of clinical trials of pharmaceutical products, UMIN-CTR is expected to be the main clinical trial registry in Japan that accepts registration of clinical trials to evaluate surgical methods, medical devices, foods, and educational programs.

UMIN-CTR

UMIN-CTR was designed originally as a clinical trial registry that conformed to the ICJMJE statement (Tab. 4). After the World Health Organization (WHO) started their International Clinical Trials Registry Platform (ICTRP) project [33, 34], in efforts to follow ICMJE, UMIN-CTR was redesigned to meet the WHO requirements for clinical trial registries. UMIN-CTR began in June 2005.

Approximately 340 clinical trials have been registered with UMIN-CTR as of February 2006. The registry provides information on clinical trials conducted in Japan, in both Japanese and English. Information in both languages is provided by the organization that submits the information at the time of trial registration. Study information in Japanese is required currently; however, information submitted in English will be required for registration of clinical trials conducted outside Japan in the future.

Table 4. UMIN-CTR: Roles and responsibilities

- Operation initiated in June 2005
- Offers information on clinical trials conducted in Japan; information offered in Japanese and English
- Designed to accommodate the requirements of International Committee of Medical Journal Editors and World Health Organization
- Acceptable registry for International Committee of Medical Journal Editors
- Sender of information is contacted every 6 months to ensure correctness of the registered information
- Ethics committee/institutional review board that approved the trial is contacted every 6 months to monitor the trial progress
- Investigator-sponsored trials not for NDA are primarily registered

Clinical trial information registered at UMN-CTR can be updated by the sender of the information at any time. The information sender is requested to confirm the data every 6 months. The primary EC/IRB that approved the clinical trial is also contacted to inquire if the trial is active.

UMIN-CTR was recognized as an acceptable registry by ICMJE in January 2006. UMIN-CTR will closely monitor the progress of the ICTRP project of the WHO with the intent of becoming a WHO member registry.

Issues in clinical trial registration in Japan

As clinical trial registration has started, some issues have been identified concerning clinical trial registration and publication of trial results through our experience.

Necessity of mandatory registration
An estimated 2700 clinical trials are newly conducted in Japan every year. Although the number of registrations at the three registries has been increasing gradually, as of February 2006,

only about 550 trials were registered. Registration of all trials is needed to guard against publication bias. To make voluntary registration mandatory, it is necessary to make registration a requirement for:

- Initiating a clinical trial for an NDA
- Receiving public research grants for investigator-sponsored trials not for an NDA
- Receiving research grants from various organizations for investigator-sponsored clinical trials not for an NDA
- Approval of EC/IRB for investigator-sponsored clinical trials not for an NDA

Reinforcing education on clinical trials

Chalmers concluded that EC/IRB have an important role in publishing clinical trial results and that they should monitor the approved clinical trials to ensure that their results are published [35]. UMIN-CTR contacts primary EC/IRB that approve the registered clinical trials every 6 months to check the progress of relevant trials, hoping to produce the effect desired by Chalmers. Although UMIN-CTR only recently started such inquires, only about 50% of the contacted EC/IRB have responded to date. Other than the well-organized clinical trial projects, the overall quality of investigator-sponsored clinical trials not for an NDA conducted in Japan is rather low. We often see poorly developed protocols of investigator-sponsored clinical trials that use the INDICE service of the UMIN. Poorly developed protocols may make the studies difficult to complete as planned or make the results inconclusive. This finding also shows that EC/IRB members are not sufficiently capable of noticing the inadequacy of the protocol as their approval is required to use the INDICE service of the UMIN. Poor experiences of both researchers who prepare protocols and EC/IRB members who evaluate them are implicated.

It has been suggested that little opportunity to receive systematic education on clinical trials in Japan has resulted in a lack of professionals who could take charge of clinical trials [36]. Educational programs on clinical trials and other relevant areas should be developed to foster researchers capable of conducting high-quality clinical trials as well as EC/IRB members who can sufficiently fulfill their responsibilities.

Future of Japanese clinical trial registration scheme

With its interest in issues concerning clinical trial registration, MHLW has granted a study group in which UMIN, JAPIC, JMA, and the National Institute of Public Health are participating. The group hopes to develop a framework and to discuss issues of clinical trial registration in Japan.

The goal is to have, by 2007, a Web site that will allow:
- Cross-sectional search of clinical trial information registered at UMIN-CTR, JapicCTI, and JMA registries in Japanese
- Easy to understand clinical trial information for the public.
- Participation in WHO member registry, with the possibility of global transmission of information on Japanese clinical trials in line with the concept of WHO ICTRP.

We hope that the issues surrounding Japanese clinical trial registration will be discussed, the needed actions will be incorporated into the framework, and that more information on high-quality clinical trials will be transmitted from Japan to the world.

References

1   Kiuchi T, Igarashi T (2004) UMIN – Current status and future perspectives. *Medinfo* 11: 1068–1072

2   University Hospital Medical Information Network. UMIN Clinical Trial Registry. Available at http://www.umin.ac.jp/ctr/index.htm (accessed 17 February 2006)

3   Standards for the Conduct of Clinical Trials on Drugs. The Ministry of Health and Welfare ordinance No. 28; March 27, 1997

4   Standards for the Conduct of Clinical Trials on Medical Devices. The Ministry of Health, Labour and Welfare ordinance No. 36; March 23, 2005

5   International Conference on Harmonization. ICH Harmonized Tripartite Guideline. Guideline for Good Clinical Practice. 1 May 1996

6   Ethical Guidelines for Clinical Studies. The Ministry of Health, Labour and Welfare announcement No. 255; July 16, 2003; revised by The Ministry of Health, Labour and Welfare announcement No. 459; December 28, 2004

7   Ethical Guidelines for Epidemiological Studies. The Ministry of Education, Culture, Sports, Science and Technology and the Ministry of Health, Labour and Welfare announcement No. 2, June 17, 2002, revised by The Ministry of Education, Culture, Sports, Science and Technology and The Ministry of Health, Labour and Welfare announcement No. 1, December 28, 2004

8   Guidelines for Clinical Studies on Gene Therapies. The Ministry of Education, Culture, Sports, Science and Technology, and The Ministry of Health, Labour and Welfare announcement No. 1, March 27, 2002, revised by The Ministry of Education, Culture, Sports, Science and Technology and The Ministry of Health, Labour and Welfare announcement No. 2, December 28, 2004

9   The Pharmaceutical Affairs Law. Law No. 145, August 10, 1960

10 The Enforcement Regulation of Pharmaceutical Affairs Law. The Ministry of Health and Welfare ordinance No. 1, February 1, 1961

11 Outline of Manufacturing Approval for Pharmaceutical Products. The Pharmaceutical Affairs Bureau notification No 645, September 13, 1967

12 Standards for the Conduct of Clinical Trials on Drugs. The Pharmaceutical Affairs Bureau notification No. 874, October 2, 1989

13 The Ministry of Health, Labour and Welfare. Yakumu Koho. Tokyo, Yakumu Koho Sha (in Japanese)

14 Tsutani K (2003) Evidence wo Shiraberu – Sisutematikku Rebyu no Genjo (Examine evidences – present situation of systematic review). *Rinsho Yakuri* (*Jpn J Clin Pharmacol Ther*) 34: 210–216 (in Japanese)

15 Tsutani K, Kaneko Y, Nakayama T et al (2002) Nihon de wa Maitsuki Yaku 70 Pen no RCT ga Houkoku Sarete Iru (70 randomized clinical trials are estimated to be reported monthly in Japan). *Rinsho Yakuri (Jpn J Clin Pharmacol Ther)* 33: 273S–274S (in Japanese)

16 Japan Hand Search & Electronic Search Society. Available at: http://jhes. umin.ac.jp (Accessed 10 March 2006)

17 Easterbrook PJ, Berline JA, Gopalan R, Matthews DR (1991) Publication bias in clinical research. *Lancet* 337: 867–872

18 Stem JM, Simes RJ (1997) Publication bias: evidence of delayed publication in a cohort study of clinical research projects. *Br Med J* 315: 640–645

19 Decullier E, Lheritier V, Chapuis F (2005) Fate of biomedical research protocols and publication bias in France: retrospective cohort study. *Br Med J* 331: 19–22

20 Pharmaceuticals and Medical devises Agency. Available at www.info. pmda.go.jp (Accessed 8 February 2006)

21 New Drug Application. Pharmaceutical and Food Safety Bureau notification No. 481, April 8, 1999

22 Narukawa M (2000) Shinyaku no Shonin Shinsa Katei no Tomeika to "Shinyaku Shonin Johoshu" ni Tsuite (Transparency of new drug review process and "New Drug Approval Package" by MHW, Japan. *Rinsho Hyoka (Clin Eval)* 27: 519-527 (in Japanese)

23 Sunaga M (2005) Publisher no Tachiba kara (From publisher's standpoint). *Rinsho Iyaku (J Clin Ther Med)* 1138: 114S (in Japanese)

24 Beppu H (1999) Japan's loss of leadership role in access to drug data. *Lancet* 353: 1992

25 Pharmaceutical Research and Manufacturers of America. Joint position on the disclosure of clinical trial information via clinical trial registries and databases. Available at: www.phrma.org/publications/policy//admin/2005-01-06.1113.PDF (Accessed 8 February 2006)

26 Kaneko Y, Tsutani K, Nakayama T et al (2002) 2000 Nen ni Nihon de Hokoku Sareta RCT no Naiyo (Analysis of contents of randomized controlled trials reported in Japan in 2000). *Rinsho Yakuri (Jpn J Clin Pharmacol Ther)* 33: 323S–324S (in Japanese)

27 Tsutani K, Hirose M, Kurihara C et al. Handsearch and electronic search projects in Japan. 8th International Cochrane Colloquium. 25–29 October 2000, Cap Town, South Africa (page 218)

28 DeAngeles CD, Drazen JM, Frizelle FA et al (2004) Clinical Trial registration; a statement from the International Committee of Medical Journal Editors. *JAMA* 292: 1363–1364

29 Translational Research Informatics Center. Center Information Japan. Available at: http://cancerinfo.tri-kobe.org/clinicaltrials/index.html (Accessed 17 February 2006)

30 Japan Clinical Oncology Group (JCOG). Research Groups. Available at: http://www.jcog.jp/STUDY_GROUP/fra_grouptop.htm (Accessed 17 February 2006)

31 Japan Pharmaceutical Information Center. Clinical Trials Information. Available at: http://www.clinicaltrials.jp/user/cte_main.jsp (Accessed 17 February 2006)

32 Center for Clinical Trials, Japan Medical Association. Clinical Trial Registration. Available at: http://dbcentre2.jmacct.med.or.jp/ctrialr (Accessed 17 February 2006)

33 World Health Organization. International Clinical Trials Registry Platform. Available at: http://www.who.int/ictrp/en (Accessed 17 February 2006)

34 Gulmezoglu AM, Pang T, Horton R, Dickersin K (2005) WHO facilitates international standards for clinical trial registration. *Lancet* 365: 1829-1831

35 Chalmers I (1990) Underreporting research is scientific misconduct. *JAMA* 263: 1405–1408

36 Fukushima M (1995) Clinical trials in Japan. *Nat Med* 1: 12–13

Clinical Trial Registries: A Practical Guide for Sponsors and
Researchers of Medicinal Products, edited by MaryAnn Foote
© 2006 Birkhäuser Verlag Basel/Switzerland

Transparency and validity of pharmaceutical research

Alan Goldhammer

PhRMA, 950 F Street, NW, Washington, DC 20004, USA

Introduction

The United States pharmaceutical industry spent more than
US $ 38 billion during 2005 in the search for new cures. Unlike
the development and marketing of many consumer products,
bringing a new pharmaceutical product to the patients who
need it is not an overnight process. Rather, it takes years, not
months, and costs in the neighborhood of US $ 800 million
[1]. Not every drug that is synthesized in the laboratories of
a pharmaceutical company makes it through development.
In fact, the overwhelming majority have their development
discontinued by the sponsor because of toxicity or lack of efficacy. Most industry experts note that for every 10 000 potential drug candidates screened, only approximately 250 proceed
to nonclinical evaluation. Of the 250 candidates, 5 may move
into clinical studies, with only 1 candidate eventually being approved for marketing. Of the drugs that receive approval from
the United States Food and Drug Administration (FDA), only
a few realize full market potential.

Almost every day, there is a news story about the pharmaceutical industry in the United States. Some of the news
is good: a new drug for a serious or life-threatening disease is
approved for marketing. Some of the news is not good: issues

related to drug safety or drug affordability. Whether the news is good, bad, or indifferent, it is incontrovertible that the drug development process is expensive, uncertain, and has resulted in differing expectations from all parties. The critical need from the research and development process is the need for valid data that clearly demonstrate the safety and efficacy of the prospective pharmaceutical candidate.

In general, the public's expectations are straight forward. The public wants prescription drugs that work for everyone, are always safe, and are uniformly inexpensive. These expectations are not reasonable. Patients differentially respond to medicines from both a safety and efficacy perspective. Furthermore, no drug, not even over-the-counter (OTC) products, is totally with out some degree of risk. The cost factor is built into the drug development process and the limited patent life of the product, which dictates how much time the company has to recoup its research and development costs. It must also be recognized, however, that the cost of pharmaceutical products ultimately comes down when the product goes off patent and is subject to generic competition. In fact, the economic argument can be made that over the lifetime that the drug is on the market, the average cost is significantly lower than the introductory market price.

The process of drug development in the United States

The two critical issues in drug development are money and time. The pharmaceutical industry is heavily regulated. Products cannot be brought to the market until the exacting standards of the FDA have been met and the drug is licensed. While the Prescription Drug User Fee Act (PDUFA) has improved the drug review process, the amount of nonclinical and clinical work that companies need to do has increased over time. Even when the drug reaches the market, all the safety and efficacy issues are likely not fully understood.

None of the development is conducted in a vacuum. Companies meet with the FDA before initiating clinical trials and during the development process. Critical research protocols are discussed to assure that the research will provide the data needed to support a license application. All clinical research is conducted to the highest ethical standards that have international recognition under Good Clinical Practice (GCP) guidelines established by the International Conference on Harmonisation (ICH), guidelines that were developed in cooperation with drug sponsors and regulatory agencies worldwide [2]. Furthermore, various regions have laws and regulations in place to assure that each study subject is provided with meaningful and understandable information about the conduct of each clinical trial, including full disclosure of the intervention name, comparator, and indication. It is routine, and in many regions a mandatory requirement, that the clinical trial protocol be reviewed by an independent ethics board (commonly called an Institutional Review Board in the United States). Only after making this full disclosure to the prospective study subject, is voluntary and informed consent from each subject obtained. All this information is provided to the relevant regulatory authorities and the institutional review boards that oversee the protection of human subjects.

Clinical trial sites are regularly audited to ensure that the data comply with all applicable regulations and is of high quality. The FDA may also do its own audits to assure compliance with GCPs.

At the end of the clinical development process, all the data are compiled, analyzed, and formatted into a New Drug Application (NDA) for FDA review. Shortly after submission, FDA determines whether the submission is for an important new drug that offers a significant benefit. In such cases, the review period is 6 months, with periods for standard reviews being 10 months. Some have argued that these review times put pressure on FDA to approve new submissions. This position is simply wrong. First, the statutory time allocated to FDA for an NDA review is 6 months. Second, the PDUFA agreements, first set forth in 1992, call for FDA to issue an action letter by the close of the review clock [3]. The action letter could be an approval or a complete response letter to the sponsor detailing issues that need to be addressed to allow for marketing approval. This level of review effort on the part of the FDA is not negligible. Using accounting figures from the FDA [4], the Pharmaceutical Research and Manufacturers of America (PhRMA) estimates that approximately 14 person-years of effort goes into the review of an NDA, a staggering amount work. (Calculation: For the fiscal year 2003, the standard cost to review an NDA was US $ 2.1 million. Assuming an average loaded full-time employee at the FDA is US $ 150,000/year, one can approximate the review time by dividing US $ 2.1 million by US $ 0.15 million.)

Finally, as the result of agreement with the FDA, the drug sponsor may need to do further studies once the product is on the market. These studies are safety studies to answer certain questions or clinical efficacy studies to validate a surrogate end point.

Ultimately, the FDA decision is based on an assessment of the benefits and risk the drug poses. For serious and life-threatening diseases, this evaluation is quite different than for a new antihistamine for the treatment of allergic rhinitis. A patient with cancer often is willing to accept greater risks for a treatment than a patient with a seasonal allergy.

Clinical trial registration

Despite this intensive interaction with the FDA and the independent review of all the clinical data by the FDA, questions have been raised about the transparency of the research process. There are two aspects to clinical trial transparency: registration of clinical trials at the time of inception and the disclosure of clinical trial results after completion of the trial.

Registration of clinical trials provides patients and healthcare providers with information on participation in trials. If during the registration process, the trial is given a unique identifier, the results of the trial can be linked to the registry when the trial is finished and the data are analyzed. The disclosure of results provides healthcare providers (principally) and patients (secondarily) information on all clinical trials conducted by the drug sponsor, regardless of outcome and whether that information is on the FDA-approved drug label.

Registration of clinical trials in a publicly accessible database at the time patients are enrolled helps assure the transparency of the pharmaceutical research process. Assigning trials a unique registration identification number allows pharmaceutical companies (and other clinical trial sponsors) to cross reference the results of clinical trials, so that the healthcare community can be assured that the results of all clinical research are appropriately communicated. A registry of clinical trials was established at the US National Library of Medicine (NLM) as a result of the Food and Drug Administration Modernization Act (FDAMA) of 1997 [5].

Although the original legislative intent of this database was to provide patients and healthcare providers with information on clinical trials for serious and life-threatening medical conditions, the pharmaceutical industry has committed to registering all hypothesis-testing clinical trials on this registry using a basic set of data elements based on the legislation that established the registry [6].

A significant amount of discussion has occurred about the future direction of clinical trial registration. The International Committee of Medical Journal Editors (ICMJE) stated that registering trials at inception was a means of improving transparency [7]. Journals subscribing to the ICMJE principles will not consider for publication any clinical trial whose primary purpose is to affect clinical practice (e.g., phase 3 trials) that was not registered in accordance to the time lines determined by ICMJE.

Drug sponsor companies that are members of PhRMA have implemented a number of policies designed to improve the transparency of clinical research. These policies are embodied in the "Principles on Conduct of Clinical Trials and Communication of Clinical Trial Results" of the PhRMA that were issued in 2002 and updated in 2004 [8]. Since issuing those principles, the PhRMA has actively engaged other stakeholders. The 2004 update contains an extended set of questions and answers. An on-line database of clinical study results was established in October 2004 to provide a venue where drug sponsors could publicly and broadly communicate the results of those trials that might not otherwise be published in the medical literature. This database can be accessed at www.clinicalstudyresults.org. As of April 2006, the database contained information about clinical trials for more than 255 pharmaceutical products.

The issuance of PhRMA's 2002 principles outlined the pharmaceutical industry's commitment to conducting clinical trials according to the highest ethical standards with respect to the patients who participate. More importantly, the statement stressed the importance of transparency in conducting clinical trials and reporting their results.

Obtaining transparency

The principal ways of accomplishing transparency in the conduct of clinical trials and publication of their results is to communicate the results of all controlled trials done on marketed products, regardless of outcome. The PhRMA has focused on marketed products because these products can be prescribed by physicians and such trials can and do affect medical practice.

The major avenues for communicating results are through peer-reviewed articles in the scientific literature or presentations at major medical or scientific meetings. The PhRMA principles note that other means for communicating results can exist.

One of the difficulties in communicating clinical trial results is that traditionally most medical journals are not interested in publishing so-called negative studies. In addition, a presentation at a scientific meeting, even one that is national in scope, may not reach all interested medical professionals.

The PhRMA database, clinicalstudyresults.org, addresses these two key issues in communicating clinical trial results. Drug sponsors can post results of their clinical trials of marketed products. As conceived by the PhRMA, drug sponsors are requested to post those studies completed after the effective date of the PhRMA Principles (October 2002), but they are free to post studies completed before that date.

Clinicalstudyresults.org

The clinicalstudyresults.org database contains three principal data elements for each drug:
1. The FDA-approved drug label, which is of course the physi-

cian's primary source of information regarding how to optimize the benefits and risks in a treatment decision
2. A bibliography of peer-reviewed publications of clinical studies with links to the study where possible
3. Summaries of those studies that are unpublished in a peer-reviewed journal, preferably in a format recognized by the ICH

The PhRMA principles state that study summaries should be posted within a year of completion of the clinical trial unless the company is seeking publication in a peer-reviewed journal. Premature disclosure of results on the Web site could compromise publication of the results (i.e., the results could be considered to be publication of enduring materials, a standard used by journal editors to determine previous publication).

The data set element issue

The common set of data elements used for registering clinical trials has been the subject of considerable discussion. The original ICMJE policy statement discussed certain data elements that the medical journal editors believed to be important for registration [7]. Subsequently, the World Health Organization (WHO) initiated a consultation on clinical trial registration standards, which resulted in the establishment of an International Clinical Trials Registration Platform (ICTRP) [9]. A summary of the initial discussions of this process was the subject of a second editorial statement from the ICMJE [10].

The key outcome of the initial WHO consultation was a statement on the types of trials that need be registered, and a list of 20 data elements that constitute the minimum for clinical trial registration. The PhRMA strongly supports this set of data elements, and is committed to disclosing this information for "all trials that prospectively assign human participants or groups to one or more health-related interventions to evaluate the effect on health outcomes" [11]. Registration of exploratory studies not designed to influence health practice and that serve only to set the direction for future testing was not supported by the original statement from the ICMJE [7]. Such trials may be registered at the sponsor's discretion.

Drug sponsors have committed to publicly disclose all 20 data elements at trial initiation except in the infrequent case where disclosure may raise competitive concerns. In such situations, however, the remaining data fields will be publicly disclosed at a later date. The WHO has also acknowledged this situation [11], and noted specifically that 5 data elements [i.e., official scientific title of the study, intervention(s), target sample size, primary outcome, and key secondary outcome] may be regarded as sensitive for competitive reasons by the study sponsor who may wish to delay release of the information.

The following examples represent circumstances where certain fields would be considered highly proprietary during the drug development process and whose disclosure would be delayed.

1. A drug sponsor has a medication whose normal route of delivery is subcutaneous injection. The sponsor has developed a new approach that uses a device to administer the drug by inhalation. If a competitor has a similar drug in this therapeutic category, the sponsor may wish to maintain the confidentiality of the delivery technology. The sponsor would amend the title of the study so that the route of administration was not disclosed.

2. A drug sponsor is developing a new surrogate endpoint concurrent with developing a test to measure the surrogate and gain a patent on the new technology. The sponsor envi-

sions the surrogate will reduce the time it takes to get the product approved by the FDA. Therefore, the sponsor will delay disclosure of the surrogate as the primary outcome while it develops the novel approach.

3. A drug sponsor is pursuing an out-of-therapeutic-class new indication for a marketed product. Other drugs in the class may also have a clinical effect on the new indication. The sponsor will delay disclosure of the drug name until later in development.

4. A drug sponsor is developing a new formulation of a medicine that will permit weekly dosing *versus* daily dosing. The new dosing regimen will improve patient compliance; the information in early stage development is highly proprietary. The sponsor will not want to disclose the formulation to potential competitors.

Ongoing discussions have been moving in an ill-considered direction. It appears that WHO is moving toward requiring registration of all interventional trials (thus including, the early phase 1 and phase 2 studies previously excluded in agreements with industry trade associations) and open-label (single treatment cohort or noncomparator) studies [12]. The addition of 'interventional' studies to clinical trial registries adds little data to inform healthcare practice.

The inclusion of early phase trials is not in the best interests of the stakeholder community. The registration of phase 1 trials may stifle pharmaceutical innovation, resulting in fewer new products reaching the market. Phase 1 trials are largely focused on the safety of new compounds in humans; are designed as dose toleration, ADME (absorption, distribution, metabolism, and excretion) studies in healthy volunteers; and are not interventional as they do not evaluate health outcomes. These trials do not result in information that can inform clinical practice. The concern is that inclusion of these sorts of studies in any registry database will only serve to confuse, rather than inform.

Summary

It is unclear where this debate is headed and how it will be resolved. It is important to remember that patients' needs are quite different from the research community's when it comes to clinical trial registries. The patient is interested in the number of different trials, the enrollment criteria, and the geographical proximity of the trial site so that he/she can make a decision about whether the trial is of potential interest. These are a mere subset of the 20 data elements that are under discussion and have little bearing on the transparency of clinical research that the remaining data elements are designed to address.

Finally, the role of the FDA must be fully recognized for what it is, an impartial review of all the data required to support a marketing application. Those who castigate the pharmaceutical industry indirectly impugn the integrity of the FDA. Pharmaceutical industry clinical research goes forward with the full knowledge that trial design and results will be critically examined. A product will not be licensed for marketing nor will an advertising claim be permitted unless the data support it.

References

1 DiMasi JA, Hansen RW, Grabowski HG (2003) The price of innovation, new estimates of drug development costs. *J Health Econ* 22: 151–185
2 ICH. Good Clinical Practice Guidelines; Guidance Document E-6(R1). Available at: www.ich.org (Accessed 23 April 2006)

3 Prescription Drug User Fee Act. Available at: http://www.fda.gov/cder/pd-fua/default.htm (Accessed 23 April 2006)
4 Fiscal Year 2003 data. Available at: http://www.fda.gov/cder/pdufa_costs.htm (Accessed 23 April 2006)
5 United States Food and Drug Administration. Section 113 Food and Drug Administration Modernization Act of 1997. Available at: http://lhncbc.nlm.nih.gov/clin/113.html (Accessed 8 February 2006)
6 Pharmaceutical Research and Manufacturers of America. Principles for registering clinical trials. Available at: http://www.phrma.org/publications/policy_papers/phrma_clinical_trial_registry_proposal (Accessed 23 April 2006)
7 DeAngelis C, Drazen JM, Frizelle FA et al (2004) Clinical trial registration: A statement from the International Committee of Medical Journal Editors. *N Engl J Med* 351: 1250–1251
8 Pharmaceutical Research and Manufacturers of America. Principles on conduct of clinical trials and communication of clinical trial results. Available at : http://www.phrma.org/clinical_trials (Accessed 23 April 2006)
9 World Health Organisation. International Clinical Trial Registry Platform. Available at www.who.int/icrtp/en (Accessed 23 April 2006)
10 DeAngelis C, Drazen JM, Frizelle FA et al (2005) Is this clinical trial fully registered? A statement from the International Committee of Medical Journal Editors. *N Engl J Med* 352: 2436-2438
11 World Health Organisation. WHO Consultation. Available at: http://www.who.int/ictrp/news/ictrp_sag_meeting_april2005_conclusions.pdf   (Accessed 23 April 2006)
12 World Health Organisation. Comment solicitation. Available at: http://www.who.int/ictrp/comments4/en/print.html (Accessed 23 April 2006)

A project management approach to the planning and execution of clinical trial registries

Mark Jungemann

Pharmaceutical Project Management, Eli Lilly and Company, Lilly Corporate
Center, Indianapolis, IN 46285, USA

Introduction

The creation and the maintenance of a clinical trial registry is a very complex process. Creating a clinical trial registry requires both an awareness of the environment external to the sponsor and the ability to direct internal resources effectively. Various regulations, including FDA Modernization Act (FDA-MA) section 113 [1], guidelines from the Pharmaceutical Research and Manufacturers of America (PhRMA) [2] and the World Health Organization (WHO) [3], and publicly stated opinions address potential elements of a clinical trial registry. Currently, no single global standard exists for a clinical trial registry and the positions and requirements of the various stakeholders continue to change. For sponsors, the numerous and frequently changing regulations, guidelines, and opinions create a challenging external environment that is not consistent from country to country. The impact of this environment is felt across many of the sponsor's internal departments. At Eli Lilly, participating departments include Medical, Scientific Information, Regulatory Affairs, Corporate Communications, Clinical Operations, and Information Technology. Because

Lilly conducts clinical trials globally, the company affiliates also participated in creating the registry.

The complexity created by both internal and external forces should compel the prudent sponsor to apply a systematic approach to the creation and maintenance of a clinical trial registry. A project management approach was selected by Lilly to ensure that the Lilly registry would be well defined, have consistently high quality, and be delivered in time to meet external requirements.

This chapter discusses the fundamentals of project management on a phase-by-phase basis and provides examples of how Lilly applied the basics of project management to its clinical trial registry project.

Defining a project

A project is a temporary endeavor to create a unique product, service, or result [4]. If a registry is intended to be a long-term solution to the issue of transparency in clinical trials, how can this "temporary endeavor" meet the definition of a project? At Lilly, the clinical trial registry project was designed to provide a business process, communication, training, information technology support, and on-going implementation of the public disclosure of the results of all trials completed for products that obtained marketing approval for which Lilly was the sponsor. Given this purpose, the project was designed to create a long-term process that could be implemented consistently across functions and countries. The project approach was selected because the process to be developed had both an immediate and a longer-term use. The end result of this temporary endeavor was both the process and a unique product, namely, a company-specific registry, part of the criteria for a project.

Lilly's approach

Lilly's policy on public data disclosure [5] is the document that set the standards for the registry. Lilly's policy includes both near-term and longer-term requirements. A project approach allowed Lilly to meet its near-term requirements for building its registry and allowed Lilly to build a longer-term process for keeping its registry current. The project approach also allowed Lilly to use a larger number of temporary resources as it trained a few individuals who would be retained to run the registry process for the long term.

Project phases

A project can be broken into phases or project management process groups: initiation, planning, execution, monitoring and control, and closeout [4].

Initiation
The first phase, initiation, is often omitted when a project management approach is not used. For the Lilly registry project, initiation set the vision for the project. The vision, documented in the form of a project charter, was drafted by the project manager and approved by the project sponsor and steering committee. Optimally, a charter should be completed early in the project to minimize misunderstanding among participants and between participants and management. The project charter helps all stakeholders understand what is expected of the project, when it is to be completed, what level of quality will be required, what resources will be required to complete the project; and when the project will be considered to be completed (Tab. 1). Lilly's clinical trial project charter contained

Table 1. Sample questions to guide the writing of a project charter

- What is the project scope and what is the project team's responsibility?
- What does a high-quality result look like?
- How much volume (number of studies) is expected?
- How many people will be needed and what skills will they need?
- How will the team be supported by management and other functions?
- What will the team do if the requirements (often external environment) change?
- How will participants be rewarded?
- Who are the decision makers?
- How will the team communicate the plan to all who need to know?

Table 2. Sponsors of Lilly's clinical trial registry project used nine criteria to measure success

- I will be assertive in determining project direction (scope)
- I will ensure the project deliverables are things I support
- I have clearly communicated what I expect from the project
- I will 'Inspect What I Expect' (ask for progress reports)
- I will be accountable for cross-functional project support
- I will take a stand on prioritizing quality, time, cost, or scope
- I will coach the project leader and team members for improvement
- I will remove obstacles at the appropriate level
- I will provide communication and feedback to the team

nine sections (Project History; Background and Purpose; Project Benefits; Assumptions, Constraints and Dependencies; Project Scope and High-Level Schedule; Quality Plan; Issue Escalation and Change Management; Risk Management Plan; and Team Roles and Responsibilities).

A successful project usually requires active sponsorship. While some project sponsors simply provide the resources and wait for project completion, the Lilly registry project sponsors were executives who directly helped the project team. Lilly's sponsors did not manage the project, but they were active and continually involved at an appropriate level as major project decisions and milestones were reached. As part of the charter, the sponsor had set criteria by which to measure success (Tab. 2).

Table 3. Role of sponsors in early project planning stage (adapted from [6])

- Assisting the project manager in establishing the correct objectives
- Providing guidance for the project manager in organizing and staffing
- Explaining to the project manager the environment and political factors that could influence the project's execution
- Establishing the priority for the project
- Providing guidance for the establishment of policies and procedures by which to govern the project
- Functioning as the executive-client contact point

Table 4. Project planning helps to determine what must be accomplished to meet goals. During planning, some actions must be taken and some questions answered

- Finalize the scope of the project and set expectations in a concise statement
- Divide the work and create a detailed schedule
- Determine what needs to be done (specific tasks with actions to complete defined project deliverables)
- Assign responsibility for tasks, i.e., who does what, how is it organized, and what resources are required?
- Determine when tasks and deliverables due
- Sequence the tasks
- Describe what 'good' looks like (quality planning)
- Document actions to take if something goes wrong by conducting risk planning and issue escalation

The project sponsors at Lilly were particularly involved during the initiation phase of the project when the scope, policies, and resources were defined in keeping with the idea that during the initiation phase, project sponsors should have an active role [6] (Tab. 3).

Planning

Project planning is the second phase in managing a project; it takes its direction from the project charter and defines clearly how the goal will be achieved (Tab. 4).

Work streams

Dividing the work helps make describing project tasks easier and helps the scheduling process. Lilly's clinical trial registry project was divided into eight parts (subprojects) or work streams (Tab. 5). Work streams operated under several rules. Each work stream leader was accountable for results of his or her work stream and was responsible for various tasks, such as staffing; determining subtasks, duration, logic, and responsible person; defining the quality of the deliverables in writing and obtaining agreement with the project leader; providing periodic updates of the work stream schedule and the percentage of work completed; conducting risk identification and quantification (e.g., impact, cost, how likely); presenting periodic updates for project sponsors; keeping the project leader informed of issues and suggested resolution; maintaining documentation during the project; and completing project closeout documentation when the work stream was completed.

Lilly's experience with the work stream approach was very positive. Since the project was complex and the external environment continued to shift during every project phase, the work stream approach helped the project team maintain its focus and adapt quickly. The accountability of each work stream leader was one key to this success. Each work stream leader was an expert in the field most closely related to the work stream, and he or she understood the specific requirements for the work stream, allowing the project manager to integrate across the work streams to ensure that the overall project deliverables were met. Work stream leaders were responsible for developing a schedule for the deliverables of their work stream and developing a communication plan. Because most work streams had a global membership, relatively few face-to-face meetings were held, and communication was largely by e-mail and teleconference.

Table 5. Dividing the work into manageable parts. The Lilly clinical trial project had eight work streams

---

- Completed trials (summaries and citations for completed clinical studies)
- Initiated and ongoing trials (data required within 21 days of first patient on study)
- Processes (job aides and standard operating procedures)
- Training (global training development and delivery)
- Audit (internal and third party)
- Quality (define what 'good' looks like and document quality assurance processes)
- Communication (internal and external communication)
- Information technology (system links, automation)

Project schedule

A project schedule ensures that all tasks are completed on time (Fig. 1). A schedule is developed through a series of steps, including definition of the activity, sequencing of the activity, estimating resources required, estimating the time needed, and developing a schedule [4].

Quality standards

During planning, the team identified ways to achieve specific quality standards. In the early stages of the project, the appropriate quality standard for the project deliverable was determined and documented. Using both external and internal standards, a quality work stream group developed quality standards that took the form of templates and quality control check lists. These standards were used by members of the training work stream to train those entering data for each clinical trial to be posted on the Web site so it met the Lilly quality standard. Quality control processes were developed to leverage the expertise in the international affiliates, and a central group was responsible for ensuring that the content posted on the Web site was not promotional in nature. The

Lilly's Project Schedule – used as a tool

Task name	% Complete	Responsible	Duration	Predecessors	Start	Finish
Project start	100 %	PM Lead	0 days		Wed 1/26/05	Wed 1/26/05
INITIATED AND ONGOING TRIALS	36 %	WSL 1	165 days		Wed 1/26/05	Tue 9/13/05
COMMUNICATION	99 %	WSL 2	35 days		Wed 1/26/05	Tue 3/15/05
QUALITY	94 %	WSL 3	126 days		Wed 1/26/05	Wed 7/20/05
PROCESS	45 %	WSL 4	240 days		Wed 1/26/05	Tue 12/27/05
PROCESS: Develop job aides for writing, reviewing and approving CTR summaries	100 %	Smith	15 days	129	Wed 3/9/05	Tue 3/29/05
PROCESS Develop job aids for determining eligible trials	100 %	Johnson	2 days	16	Wed 1/26/05	Thu 1/27/05
PROCESS: Develop appropriate SOPs for CTR Summaries	85 %	Johnson	45 days	49,50,51	Wed 3/30/05	Tue 5/31/05
PROCESS: Revise DocMan CTR Summary Template	100 %	Smith	15 days	129	Wed 3/9/05	Tue 3/29/05
PROCESS: Develop process to check Publication Status For Study	100 %	Smith	5 days	129	Wed 3/9/05	Tue 3/15/05
PROCESS: Develop process to QC Final CSR And Summary – part of job aid	100 %	Johnson	10 days	129	Wed 3/9/05	Tue 3/22/05
PROCESS: Decision Point; Storage Tracking	100 %	Johnson	5 days	129,16	Wed 3/9/05	Tue 3/15/05
PROCESS DELIVERABLE: Deliver approved Summary SOP modifications	100 %	Johnson	0 days	49,52,54,58	Tue 12/27/05	Tue 12/27/05

Figure 1. An example of a hypothetical Eli Lilly's project. CTR: clinical trial registry; PM: project management; QC: quality control; SOP: standard operating procedures; WSL: work stream leader.

quality standards that were created and tested during the project were used during implementation of the long-term process.

Communication

Communications planning included understanding the communication requirements for each key stakeholder, including the type of communication that would be most useful and frequency of communication. The members of the communications work stream developed a plan for routine and special communications both internally and externally.

Training

The training work stream group developed and delivered training to all levels of the organization globally. Subject matter experts from many functions and departments worked with project team resources to develop training modules that were specific to the deliverable required. The training work stream applied a number of rules for its work (Tab. 6). Modules were developed for writing the summaries of completed clinical trials, reviewing and approving the summaries, entering data for initiated and ongoing trials, and reviewing and approving those entries. Training was given to staff in the United States in person and to staff in other countries by Web cast. Computer-based training modules were subsequently developed and made available on a Web-based system.

Issue escalation and change management

Issue escalation and change management were handled by the project team using a simple set of rules that were documented in the project charter. The project leader was allowed to use discretion regarding planning and execution of tasks as long as the overall project deliverables were not effected. Issues that

Table 6. Rules for the training work stream

- Training will be conducted across major markets globally
- Training package must be piloted and revised before full delivery begins
- Trainers must be qualified before they deliver training
- Trainees will receive job aids and copies of Standard Operating Procedures
- Training must be completed before 1 July 2005
- Training will be designed to create behavior change
- Training will include a detailed understanding of how to write summaries of completed trials without promotional language or intent

required changes in scope, cost, time or quality were escalated to the project sponsor for approval.

Change management was purposely less formal than what would be expected in projects of the size and complexity of Lilly's clinical trial registry project, and it could be argued that a more formal process would be appropriate. The rate of external change was high, so rather than a formal process, a continuous dialogue was developed between the project team, functional management, and sponsors that helped decisions proceed in a timely fashion and allowed the team to move forward despite the rapidly changing external environment.

Execution

Project execution is not just accomplishment of the assigned tasks according to the project schedule. The team must be fully formed, and the quality standards must be maintained as the work is accomplished. Communication planning is executed so that the broad group affected by the project is kept informed. Progress towards overall completion must be monitored and reported. Scope, quality, and time are all important factors in execution of the project plan. One of the tools available to a team during project execution is risk management, and specifically, the use of signposts, an indictor of future events. Signposts help the team know when to act and they help the project team know when it has reached the point at which they need to know enough to decide on a direction (e.g., acting on a contingency plan). Signpost monitoring also requires the team to know what information is necessary to make a decision. Little is accomplished if the project team is forced to abandon the primary plan only to find that execution of the related contingency plan was started too late to realize its benefits.

The entire project team took responsibility for balancing risk (Fig. 2). As part of risk management, the team identified certain signs that all participants were to pay attention to during execution of the project.

What if Something Goes Wrong?

Risk Description	Impact	How Likely	Signposts
Team resources unavailable to write summaries	High	Med - High	April 15, 2005 for summary writing to begin
Quality of summaries does not meet requirements	High	Low	May 15, 2005 for training to begin
External factors force changes in Initiated and Ongoing Trials – cannot post non - IND studies due to changes in required fields	Med	High	April 25, 2005 for change to clinicaltrials.gov required fields

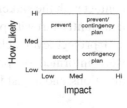

Figure 2. Consequences if something goes wrong. IND: investigational new drug.

Monitoring and control

Although monitoring and control processes were used during all phases, these processes were particularly useful in the execution phase. Throughout execution, the team reported progress against deliverables. The team routinely updated the project schedule, as tasks were completed, and identified areas of concern. The change control process was used continuously during project execution, because the scope of the project changed more than once due to changes in external factors. The project team and sponsors discussed and accepted the scope changes, and if needed, a new schedule and resource needs estimate were prepared. As work was drafted, the quality control processes were enacted, allowing the final work to be approved by predetermined individuals who had received appropriate training. Monitoring and control was useful as the team worked to accomplish the stated deliverables. If a particular deliverable was determined to be in jeopardy, project resources were redirected to ensure success. This redirection was accomplished with full knowledge of the effect on other project work.

Monitoring and control allowed the project team to stay in contact with the stakeholders. The project schedule was shown as a Gantt chart with progress bars for each major work stream and project-level deliverable. From this view, stakeholders were informed on progress and could ask relevant, timely questions, and adjust functional resources as necessary.

Closeout

In this step, each aspect of the project plan is verified for completeness and, possibly more importantly, the project deliverables are documented. The client must approve the final deliverable and verify that the final product achieves the quality standards set at the beginning of the project. During project closeout, the Lilly project team was grateful that they had written a project charter and had taken time to plan the project. Closeout documentation was done first by each work stream and then by the project leader to integrate the deliverables. Each work stream leader wrote a closeout document that detailed the planned work stream deliverables, actual work stream deliverables, timing of completion, and relevant quality standards used by the work stream. Examples of all the project work were attached to the closeout documentation. A final step in the closeout process was to document lessons learned. The Lilly team identified several lessons learned (Tab. 7).

Table 7. Lessons learned from the Lilly clinical trial registry project

- Obtain cross-functional support before asking for staff
- Create a cross-functional project team; try to have some staff fully dedicated to the project
- Divide the work into manageable parts
- Obtain agreement before starting so it is clear what each deliverable will contain
- Make good use of pilot activities before full implementation
- Document risks, mitigation, and contingencies before they occur
- Treat the project like a project; use a project management approach
- Understand that the project plan is more than a schedule
- Create and follow decision processes
- Spread the work so many staff are involved
- Make a decision once and stick to it
- Develop processes and templates so work is consistent
- Communicate frequently at all levels
- Obtain the services of a project manager to lead the team
- Engage sponsors and the steering committee so they can support project team efforts

Conclusions

The Lilly clinical trial registry project was a complex set of activities. Timelines were short, the external environment changed rapidly, and the quality standard was challenging. The work done by the team reflected a sense of true dedication. The project approach used by the team needed to be effective because there was no time to fail and begin again. External environmental shifts forced a number of changes in direction, but the project team felt strongly that the disruption by external forces was minimized because of robust project initiation and planning processes. Because of detailed planning processes, the team had alternatives when barriers occurred towards completion of the primary plan. An alternative to the project management approach would be to simply identify the tasks necessary to accomplish the work and begin to accomplish each task – the "just get it done" approach. Given the short period allowed for completion, this latter approach might seem to be the best approach. Lilly, however, believed that while this approach might get work started more quickly, the result would not have been satisfactory. Project management allowed the team to know what they needed to accomplish, when it should be accomplished, who should do the work, and which quality standard should be followed. Project management also allowed the team to recognize when its work was finished, and it allowed the team to react to changes with flexibility because many of the changes had been anticipated. The project management approach resulted in on-time delivery, carry-over processes for the long-term, and consistent high-quality work. These results speak to the benefit of using a project management approach for situations where change is likely, results are critical, resources are few, and time is short.

References

1   FDAMA Section 113: Status Report on Implementation; Executive Summary; www.fda.gov/oashi/clinicaltrials/section113/113report (Accessed 30 January 2006)
2   PhRMA Clinical Trial Registry Proposal; www.phrma.org/publications/policy/06.01.2005.1111.cfm (Accessed 30 January 2006)

3 WHO Technical Council on Clinical Trial Registries Standards; www.who.int/ictrp/news/ictrp_sag_meeting_april2005_conclusions.pdf (Accessed 30 January 2006)

4 *A Guide to the Project Management Body of Knowledge* (PMBOK Guide) Third Edition. Project Management Institute, Four Campus Boulevard, Newtown Square, PA 19073-3299 USA

5 Principles of Medical Research, Clinical Trial Registry; www.lillytrials.com (Accessed 30 January 2006)

6 Kerzner H (1998) *Project Management. A Systems Approach to Planning, Scheduling, and Controlling*, Sixth Edition. John Wiley & Sons, Inc., New York

Clinical Trial Registries: A Practical Guide for Sponsors and
Researchers of Medicinal Products, edited by MaryAnn Foote
© 2006 Birkhäuser Verlag Basel/Switzerland

Biopharmaceutical companies tackle clinical trial transparency

Dan McDonald and Steve Zisson

Thomson CenterWatch, 22 Thomson Place, 47F1, Boston, MA 02210, USA

Introduction

Pharmaceutical companies, under growing public and legislative pressure to provide more information about clinical research on new drugs and to restore public trust in the industry, have taken significant steps towards clinical trials transparency, spending an estimated US $ 89 million to post active and completed trials on publicly accessible Web sites (Fig. 1). In 2005 Thomson CenterWatch conducted an analysis of industries' efforts to improve access to information about their active and completed clinical trials; the results are reported here.

As part of industry-wide efforts to expand access to clinical trial data, many major pharmaceutical companies have published the results of hundreds of hypothesis-testing studies completed on approximately 200 marketed drugs through a centralized Web site created by the industry trade group Pharmaceutical Research and Manufacturers of America (PhRMA) [1]. In addition, several major pharmaceutical companies, including Eli Lilly, GlaxoSmithKline, AstraZeneca, and Roche, have invested considerable resources to establish trial registries on their company Web sites. While pharmaceutical companies are legally required to register trials for products treating serious or life-threatening diseases on a National In-

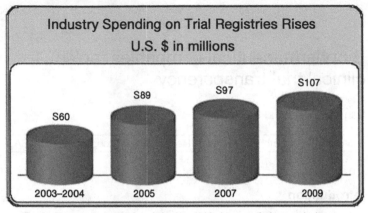

Source: Thomson CenterWatch analysis of top 15 biopharmaceutical companies by
products in development, 2005

Figure 1. Pharmaceutical industry spending on clinical trial registries, given in US $.

stitutes of Health Web site (www.clinicaltrials.gov), some companies have begun voluntarily to post clinical trials for other
types of studies on this Web site, on company Web pages, or on
commercial registries such as Thomson CenterWatch's Clinical Trials Listing Service (www.centerwatch.org).

We found that the number of studies listed on public online
registries grew by 51% from July 2004 to July 2005; and during the past 5 years, the average number of active listings has
grown from 12,300 to 32,000 per month (Fig. 2).

Industry-wide efforts to improve access to clinical trials information are increasing, yet problems remain. In our study,
we found that while 15 major pharmaceutical companies had
started aggressive transparency initiatives, similar disclosure
efforts from 35 other pharmaceutical companies, which represent more than 40% of the pipeline generated by the top
50 companies, has lagged. Existing initiatives for clinical trials transparency are fragmented, with no agreement on com-

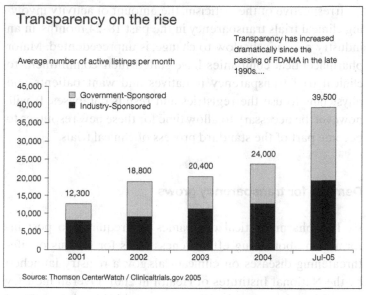

Figure 2. Transparency in reporting clinical trials has increased since the passage of FDAMA 113. The average number of listings per month is illustrated.

mon standards or scope of disclosed information. Reporting of clinical trials results remains voluntary and public criticism of the pharmaceutical industry's efforts to publicize its clinical research ranges from charges of hiding data to publishing information that is not useful for patients. The pharmaceutical industry itself admits the large number of study summaries available on the Internet can be overwhelming for patients who do not understand the clinical trials process (personal communication, M Ridge, Wyeth). Many listings found on registries such as clinicaltrials.gov appear simply to be highlights from protocols, filled with highly technical medical information that the average patient probably cannot understand.

Irrespective of the criticism, the amount of activity involving clinical trials transparency in the past 18–24 months, in an industry notoriously slow to change, is unprecedented. Major pharmaceutical companies have stated their commitment to clinical trial transparency initiatives and want patients and physicians to use the registries and results databases. It will, however, be necessary to allow time for these new resources to become part of the standard process of clinical trials.

Demand for transparency grows

By law, pharmaceutical companies are required to post information about drug effectiveness trials for serious or life-threatening diseases on clinicaltrials.gov, a registry launched by the National Institutes of Health in 2001. Overall industry compliance, however, had been poor with less than 50% of serious or life-threatening clinical trials posted during the first 3.5 years of the site [2].

For years, academic research centers and physicians have asked pharmaceutical companies for more information about new therapies and clinical trials. The issue gained momentum in 2004 when New York Attorney General Eliot Spitzer sued GlaxoSmithKline for allegedly hiding results from trials showing that its antidepressant Paxil might increase suicidal thoughts in children and adolescents [3].

GlaxoSmithKline, which settled the suit for US $ 2.5 million without admitting wrongdoing, in September 2004, launched the first public clinical trial register on its company Web site to post results and protocol information from its sponsored trials of marketed therapies. Other companies, including Roche, Eli Lilly, and AstraZeneca, soon followed with their own on-line registries.

In September 2005, PhRMA announced plans to create a central site for pharmaceutical companies to post voluntarily results from clinical trials on drugs that had received marketing approval [1]. Moreover, in April 2005, Thomson Center-Watch launched its own registry and results database on centerwatch.com and has assisted companies with complying with FDAMA and International Committee of Medical Journal Editors (ICMJE) guidelines.

In early 2005, legislation called the Fair Access to Clinical Trials Act (FACT Act) was introduced in the United States Senate that would require the reporting of all clinical trial results and expand the current mandate for trial registration [4]. The FACT Act helped focus the pharmaceutical industry's attention on the issue of clinical trials transparency. More effective, however, was the announcement by ICMJE in September 2004 that their journals would consider publication of clinical trial results only if the trial had been posted to a publicly accessible registry before 13 September 2005. In conjunction with the publication of the ICMJE statement, clinicaltrials.gov expanded its scope and modified data fields to accommodate the registration requirements of the ICMJE.

Overall, the pharmaceutical industry has adhered to the idea of voluntarily registering clinical trials on publicly accessible sites. In January 2004, the International Federation of Pharmaceutical Manufacturers and Associations (IFPMA) and three trade associations, including PhRMA, issued a joint position on the disclosure of clinical trial information, which recommended that pharmaceutical companies voluntarily submit information about all non-exploratory industry-sponsored trials to publicly accessible registries such as clinicaltrials.gov [5]. The statement also urges companies to submit summary results of industry-sponsored trials, regardless of outcome, for therapies that have received marketing approval.

Transparency increases

The ICMJE policy increased the number of ongoing trials registered [6]. Overall, companies have made significant progress
with registering their data in a relatively short period. Our
study found that 100% of the top 15 pharmaceutical companies
are posting their active clinical trials online in some capacity.
Specifically, 93% of the companies registered their studies on
clinicaltrials.gov, while 70% of the companies posted the information on commercial registries such as centerwatch.com. In
18 months, the top 15 pharmaceutical companies have posted
more than 1300 trials (Fig. 3).

At the same time, however, efforts to post clinical trial results lag behind the progress made in registering active clinical
trials. Our study found that approximately 87% of the same 15
pharmaceutical companies post trial results online with most
companies posting results to the PhRMA site (clinicalstudyre-
sults.org).

Industry response divided

In the drive towards transparency, pharmaceutical companies
have not agreed on how much information should be revealed
about either new studies or completed studies for therapies
that have received marketing approval.

The World Health Organization's (WHO) technical advisory group developed registration standards, which have been
adopted by clinicaltrials.gov and other registries, requiring
sponsors to complete 20 data fields when registering a clinical
trial. Five of the data elements, however, have been the topic
of intense debate [7]. Some companies believe that the level
of some detail required (i.e., the official scientific title of the

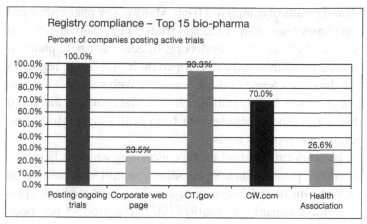

Figure 3. Registry compliance of the top 15 pharmaceutical companies: percentage of companies posting active clinical trials.

study, interventions, target sample size, primary endpoint, and secondary outcome measures) could alert competitors about the types of drugs being developed by a specific company. Major pharmaceutical companies, including GlaxoSmithKline, Eli Lilly, and Pfizer, have decided not to submit data in all 20 fields when posting certain clinical trials to protect intellectual property concerns [6]. Some companies have argued that it is not necessary to disclose these data until the therapy is available for patients in the general population. This stance has not been well received by the editors of the New England Journal of Medicine, a member of ICMJE [8].

Other pharmaceutical companies, including Wyeth and Amgen, have decided to complete the 20 data fields when registering clinical trial protocols [6]. The inconsistencies in completing the 20 data fields has removed some of the value of publicly accessible registries.

Pharmaceutical companies have chosen different paths about how to disclose information concerning both active trials and results of completed trials. Many major pharmaceutical companies register their clinical trials on clinicaltrials.gov and results on clinicalstudyresults.org. In addition, some pharmaceutical companies have developed their own online registries and databases. Some of these company Web sites post only active trials, others only publish clinical trial results of marketed drugs, and others include data from both ongoing trials and results of completed trials.

No uniform standard exists concerning what study results should be posted. GlaxoSmithKline committed to posting on its Web site the trial results of all of its marketed therapies since the company's formation [6]. If the product had been marketed in at least one country, GlaxoSmithKline posts results of all studies, from every phase, in every country. The more widespread approach adopted by pharmaceutical companies has been to post clinical study results from phase 2 to phase 4 trials.

Companies that register clinical trial protocols on clinicaltrials.gov and results for marketed products on clinicalstudyresults.org currently are unable to disclose study results from trials of therapies that do not obtain marketing approval. For example, some pharmaceutical companies have policies on the public disclosure of clinical trials and clinical trials results that call for posting information about investigational drug candidates that will no longer be developed, including new uses that will not be pursued for marketed purposes, within 2 years of ending the product development. These policies call for disclosing any medically or scientifically important information from all phase 2 or phase 3 studies of safety or efficacy that enroll patients. Because clinicaltrials.gov and clinicalstudyresults.org both lack the ability to post information without an accompanying package insert, some companies have redesigned their Web sites to include information about trials that have ended.

Eli Lilly, on the other hand, has developed a Web site that lists both clinical trial registry and results. By late 2005, the Lilly Web site included more than 100 clinical trial result summaries for 16 products for the past 10 years. In addition, the Web site includes information on more than 170 initiated and ongoing Lilly-sponsored phase 2 to phase 4 clinical trials. Lilly decided to post both registry and results on one site, in part, to avoid the appearance of incomplete disclosure. Pharmaceutical companies cannot easily hide the results of trials that have been disclosed in advance.

Other companies plan to develop Web sites that publicize both active trials and results. Bristol-Myers Squibb, for example, is developing a Web site to supplement clinicaltrials.gov and clinicalstudyresults.org for the centralized reporting of registration and results [9], and Wyeth planned to include all study postings and trial results on a redesigned Web site in 2005 (personal communication, M Ridge, Wyeth).

Roche chose another approach to disclosing trial registry and results data and has decided to publish clinical trials data through an independent host. In April 2005, Roche announced an agreement with Thomson CenterWatch to list its clinical trial protocol registry and results database [10]. Ultimately, Roche's database will include protocol information and results from all phase 2 to phase 4 clinical trials completed after 1 October 2004. Additionally, all results from phase 2 to phase 4 clinical trials for therapies marketed after 1 October 2004 will be included retrospectively. As of September 2005, Roche had posted data from 195 studies representing more than 30 medicines; 170 postings were to register clinical trials and the other 25 postings were trial results. The data will be posted to be consistent with the information disclosure principles published earlier by the European Federation of Pharmaceutical Industry Associations (EFPIA).

When deciding how to disclose clinical trials information, Roche wanted both registry and results on the same site for easier access for users. The company also wanted a global clinical trials registry and results database that was not associated with either a single country or a trade association. At the same time, Roche decided to place the registry and results database outside the company to keep the process independent and to minimize bias. Once data are transferred from Roche to Thomson CenterWatch, Roche can no longer change it, access it, or modify it without intervention from Thomson CenterWatch. Neither quality control nor hosting of the data is under the direct control of Roche.

Other companies, including Pfizer, have decided not to develop their own Web site. Pfizer has posted enrollment information for 265 ongoing studies on clinicaltrials.gov and centerwatch.com, and results from 314 studies involving 35 marketed medicines on clinicalstudyresults.org.

Resources applied

Many large pharmaceutical companies have committed significant personnel, time, and money to the issue of clinical trials transparency. Thomson CenterWatch estimates that the top 15 pharmaceutical companies spent an estimated US $ 89 million in 2004 and 2005 to disclose clinical trial information, including registering their listings in public databases and creating their own registries. This sum does not include initiatives to recruit patients.

Thomson CenterWatch analysis, based on PhRMA and company reports, estimates industry spending on registry initiatives, including development, posting, and maintenance, will continue to increase for the next 4 years, until approximately

2010. These growth projections are based on expected increases in labor and technology costs, annual contract increases, staff training costs, lack of efficiencies, and the amount of past trial data companies need to post. In the future, total spending will level as companies develop better internal workflow of these functions and complete the posting of retrospective trial data.

At a time when the pharmaceutical industry is attempting to contain headcount, our study suggests that major pharmaceutical companies have dedicated anywhere from 25 to 40 full-time equivalents, depending on the retrospective work required, to clinical transparency efforts. To produce and maintain registries, personnel are needed in areas such as project management, Web development, data entry, medical writing, regulatory, and legal. Once the backlog of trial data has been processed and systems established, large pharmaceutical companies will require 5–10 full-time equivalents to maintain the registries, depending on the volume of research.

GlaxoSmithKline has dedicated approximately 40 full-time equivalent staff to work on summaries and databases for its clinical trial results register [11]. For the initial year to 1.5 years after launch of its database, the company retrospectively generated summaries for studies that had been completed since the merger of the company in 2000. In some cases, the company posted results of studies completed before 2000 if the results were medically important. Since the introduction of the clinical trial register in September 2004, GlaxoSmithKline posted the results of more than 1100 trials representing more than 30 medicines. Once the retrospective work is finished, the company expects to have posted between 1800 and 2000 clinical trial results on its register [11]. When GlaxoSmithKline moves into registry maintenance phase in 2006, the company will need 6–10 full-time equivalents per year to cover the entire spectrum of activities, from registering the study protocols on clinicaltrials.gov to providing the results on both the company Web site and clinicalstudyresults.org.

Pfizer, which has posted information on more than 500 company-sponsored clinical trials, has also dedicated significant resources to its effort. To register their trial results, product team members were required to find resources to set up a registry of trials and results from 35 therapies (personal communication, M Berelowitz, Pfizer).

Establishing registries at pharmaceutical companies is more than posting information on a Web site. Other areas are of concern (Tab. 1).

Challenges for the future

As pharmaceutical companies move forward with providing clinical trials information to the public through registries and results databases, the industry faces many challenges: how to collect information internally and how to get it into the right format for posting; meeting staffing requirements for processing, posting and maintaining this information; guarding against competitors using detailed clinical trial information to make adjustments to their own strategies; and deciding which results, positive or negative, are statistically significant enough to report.

Another critical challenge involves explaining to the public what all the clinical trial information available on Web sites truly means. A plethora of information does not mean that the information is useful. The way in which the content is written and published, and the techniques needed to access the content can determine the success of the transparency effort.

Table 1. Other areas of concern for pharmaceutical companies

- Contracting with an external consulting company to help with project management, including assistance with developing internal processes, organizing staff, obtaining resources, and updating standard operating procedures
- Developing a company-wide training program
- Training employees worldwide to contribute to the company's Web site
- Hiring additional full-time staff to meet the ICMJE's deadline of 13 September 2005

At the same time, as the issues of transparency and increasing the public trust in the conduct of clinical research have grown in importance, so has the number of Web sites disclosing clinical trials and clinical trial results. In the United States, one state, Maine, enacted a law requiring drug manufacturers who do business in the state to disclose clinical trial data, including adverse impacts, on a publicly accessible Web site approved by the Maine Department of Health and Human Services [12]. At least nine states in the United States have introduced legislation that would require public registry and disclosure of clinical drug trials. Multiple Web sites, with different data requirements, present a problem for biopharmaceutical companies. Other issues also are important (Tab. 2).

Table 2. Issues faced by pharmaceutical companies

- Lack of internal content management systems with workflow, review, and approval functions
- Project "ownership" issues
- Fear of commitment because of regulatory uncertainty
- Concerns regarding registry proliferation
- Lack of global uniform standards for a minimal dataset
- Lack of call center or customer service center infrastructure
- Concerns about competitive intelligence
- No discernible strategic benefit

Many biopharmaceutical companies have been working with the Institute of Medicine of the WHO and with academic and industry organizations as they develop standards for registration and posting of results. In addition, the IFP-MA launched a worldwide clinical trials portal in September 2005, which is an Internet search engine that allows access to online clinical trials information, including both listings of ongoing trials and results of clinical trials. The portal can access online sources of clinical trial information posted by individual pharmaceutical company sites, sites run by third parties working on behalf of these companies, and pharmaceutical industry association resources such as clinical-studyresults.org.

For true transparency, the pharmaceutical industry needs more commitment not only from small to medium-sized companies, which often lack internal resources needed for these types of initiatives, but also from academic sites doing clinical research. Only when all stakeholders in clinical research have a common set of policies will the clinical trial process be transparent.

References

1   Pharmaceutical Research and Manufacturing Association. Pharmaceutical companies to make more information available about clinical trials. Available at http://www.phrma.org/meiaroom/press/releases/06.01.2005.1112.cfm (Accessed 13 March 2005)
2   United States Food and Drug Administration. Section 113 Food and Drug Administration Modernization Act of 1997. Available at: http://lhncbc.nlm.nih.gov/clin/113.html (Accessed 8 February 2006)
3   Office of the New York State Attorney General Eliot Spitzer. Major pharmaceutical firm concealed information. Available at http://www.oag.state.ny.us/press/2004/jun/jun2b_04.html (Accessed 17 March 2005)

4   Fair Access to Clinical Trials Act of 2005. US Senate S470. Available at
    http://olpa.od.nih.gov/tracking/109/senate_bills/session1/s-470.asp   (Ac-
    cessed 21 March 2005)
5   Pharmaceutical Research and Manufacturing Association. International
    alliance of pharmaceutical associations agrees on principles for disclosing
    information. Available at http://www.phrma.org/mediaroom/press/releas-
    es/06.01.2005.1114.cfm (Accessed 13 March 2005)
6   CenterWatch (2005) Clinical trial transparency efforts multiple. *The Cen-
    terWatch Monthly* 12(11): 1–13
7   World Health Organization. WHO International Clinical Trials Registra-
    tion Platform: Unique ID assignment. Geneva: WHO. Available at http://
    www.who.int/ictrp/commnets2/en/index.html (Accessed 23 January 2006)
8   Haug C, Gøtzsche PC, Schroeder TV (2005) Registries and registration of
    clinical trials. *N Engl J Med* 353: 2811–2812
9   McDonald D, Molnari PM (2005) Breaking the trial result disclosure log-
    jam now. *Appl Clin Trials* 11: 37–38
10 F Hoffman-La Roche. Roche clinical trial registry and results database
    launched on www.roche-trials.com. Available at www.roche.com (Ac-
    cessed 16 February 2006)
11 Metz C. Interpretation and next steps to achieve an acceptable blend of
    the position statement. Clinical Trial Registries and Registers Conference,
    Pharmaceutical Education Associates, Philadelphia, PA; September 22–23,
    2005.
12 Maine Department of Health and Human Services. Clinical Trial Regis-
    tration Law Summary. Available at http://www.maine.gov/dhhs/boh/clini-
    cal_trials.htm. (Accessed 16 February 2006)

In search of "Clinical Trial Register – Version 2.0"

Lawrence E. Liberti, Lucy Erdelac and Jean Papaj

PIA-Astrolabe Analytica, A Division of Thomson Scientific, Inc., 101 Gibraltar Rd, Suite 200, Horsham, PA 19044, USA

Introduction

Imagine a Web site key word search that results in a comprehensive, up-to-date listing of results from concluded clinical trials from around the world. Consider the knowledge to be gained from that Web site's analysis function as it generates a comparison of valid, parallel data from the trials that piqued your interest. Picture the convenience of having the comparison in an easy-to-understand table that can be printed, emailed, or saved to your computer. Think of the advantages of 24/7 access to clinically relevant patient treatment information that is timely, credible, and strongly related to trial data. Envision the possible contributions not only to your work but also to evidence-based medicine.

This clinical trial register model holds enormous promise for researchers, funders, trial sponsors, patient advocacy groups, regulatory agencies, ethical review bodies, and practicing clinicians and their patients.

How would you like to access this register in the not-too-distant future? Over the past decade, much has been written about the need for this type of clinical register. Regulatory rulings and laws have spurred a growing number of Web-based registers. Unfortunately, no single register fulfills the vision of a single-source repository of parallel data using uniform standards and a layered search function to provide user data mining, comparison and analysis.

The time is right to take the best of what currently exists and focus its content and functionality on real-world user wants and needs. An appropriate metaphor for realizing this vision is the product enhancement of computer software. After product introduction in the marketplace, software designers, expert users, and early adopters collaborate to review the product's original concept to add or refine user benefits and functionality. The updated product is denoted "Version 2.0". Since the launch of the clinical trial register concept, sufficient time has elapsed for developer organizations, and stakeholder groups to reflect on and discuss the best of what exists, and the potential of what could be. It is time for collaboration as we search for "Clinical Trial Register Version 2.0".

The difference between register and registry

Medical science media often use the words 'register' and 'registry' interchangeably. For the purpose of this chapter, we have used the prevailing conventions: Listings of information regarding the identification and participation in ongoing clinical trials are aggregated in a registry. Clinical trial results, the focus of this chapter, are aggregated in a clinical trial register.

The term 'clinical trial' has been defined in numerous references with perhaps the most comprehensive one being crafted by members of the International Committee of Medical Journal Editors (ICMJE) in a 2004 editorial [1] and subsequently updated in 2005 [2].

The ICMJE describes 'medical intervention' as meaning any intervention used to modify a health outcome. It further defines it to include drugs, surgical procedures, devices, behavioral treatments, process-of-care changes, etc [2].

The case for registering information from concluded trials

Clinical trials can be sponsored publicly or privately. Their data and findings are invaluable to medical science and form the basis for evidence-based medicine. The results of many concluded trials are never reported by sponsors, however. As a result, their findings cannot contribute to the evidence base for healthcare decision-making. Their absence widens the knowledge gap in medical research and funding, and may result in duplication of effort and wasted resources. The ultimate price associated with unreported clinical trial conclusions can occur when negative findings not promptly disseminated result in the loss of human life.

Sponsoring organizations may have many reasons for not reporting clinical trial findings. The most widely cited is the intense competitive environment surrounding many private company sponsors; dissemination of critical early findings could result in providing important clinical insights into the company's proprietary clinical development strategies. Concerns such as these are legitimate and deserve serious consideration.

Increasing stakeholder and government interest

The topic of aggregating clinical trial facts and findings into a central repository was first discussed within the medical sci-

ence community in 1986 [3]. The Web-based concept was to be
a resource through which clinical trials would be included in a
comprehensive, searchable database. Since that time, the concept has not only been embraced by the medical science community, but also patient advocacy groups, regulatory agencies
and, as a direct result, legislative bodies including the Congress of the United States of America.

The need to balance the interests of patient advocates, private clinical trial sponsors, research funders, and other groups
have resulted in two pieces of legislation currently before the
Congress of the United States of America. As of February
2006, the Fair Access to Clinical Trials Act (FACT Act) and
the Clinical Research Act are in committee. Both Acts promote the use of public registers and/or registries.

The FACT Act (S.470 and HR3196, and introduced as HR
5252 in the previous congressional session) amends the Public
Health Service Act to expand the scope of information required
for the databank on clinical trials of drugs, and for other purposes. The FACT Act requires that a clinical trial be registered
before its commencement, stipulates the trial facts that must
be publicly disclosed in a manner that is "readily accessed and
easily understood by members of the general population".

The Clinical Research Act of 2005(S 1543 and HR 2308)
provides for clinical research support grants, clinical research
infrastructure grants, and a demonstration program on partner-
ships in clinical research, and for other purposes. One of the
Act's requirements is that the Director of the National Institutes of Health (NIH) is to award clinical investigator advance-
ment grants to eligible academic health centers to support the
translation of basic science to patient care by implementing and
conducting all aspects of their clinical research mission. In addition, it also requires the Director to award clinical research infrastructure grants to eligible academic health centers to foster

Table 1. United States pharmaceutical company clinical trial registers

Company	Clinical trial register URL
Bayer Healthcare	http://www.bayerhealthcare.com/index.php?id=222&L=2
Eli Lilly and Company	http://www.lillytrials.com/
GlaxoSmithKline	http://ctr.gsk.co.uk/welcome.asp
Hoffman La Roche	http://www.roche-trials.com/

the use of information technology to facilitate the transformation of basic research findings on disease mechanisms into the development of new methodologies for diagnosis, therapy, and prevention. In addition, there are similar types of legislative initiatives affecting clinical trials registers at the state levels.

Despite early requirements for a few companies to present their clinical findings in a Web-based format (prompted by legal settlements with various State Attorneys General), increasing government interest in promoting trial data information that has the potential to benefit patient treatment has provided the impetus for repository development and the growth of both registers and registries. A large number of clinical trials are conducted by pharmaceutical companies; not all provide their results on their company Web sites, but rather publish them on consolidated Web sites such as clinicaltrials.gov or clinicalstudyresults.org. Table 1 provides a list of major pharmaceutical companies that publish proprietary clinical trial registers.

Highlights of key websites that aggregate clinical trial data

ClinicalStudyResults.org

Government interest on behalf of patients and clinicians encouraged pharmaceutical companies to make concluded clinical trial data available to the public. Collectively, through the Pharmaceutical Research and Manufacturers of America (PhRMA) they created a central, Web-based repository for clinical study results in a reader-friendly, standardized format. The stated goal for clinicalstudyResults.org is to make clinical study results for United States-marketed pharmaceuticals more transparent.

ClinicalTrials.gov

In 1997, the Food and Drug Administration Modernization Act (FDAMA) required the creation of a clinical trial database. In 2000, the National Library of Medicine (NLM) at the NIH announced the launch of clinicaltrials.gov. The site provides a regularly updated clinical trial registry of federally and privately supported clinical research in human volunteers.

Current controlled trials (ControlledTrials.com)

Launched by biomedical publishing organization Star Navigation Group in 1998, the site's goal is to increase the availability and promote the exchange of information about ongoing controlled trials worldwide. The site is free to users. Sponsor organizations are charged to register their clinical trials.

The Clinical Trial Registry Platform (http://www.who. int/ictrp/en/)

The platform is a project of the World Health Organization (WHO). Its main components include:
- Norms and standards on which trials to register, what information needs to be registered, who is responsible for registration, etc. (Clinical trial result reporting is anticipated as a future platform attribute.)
- A network of Member Registers that meet WHO-specified criteria for quality and acceptability.

- A coordinated process for detecting and resolving duplicate registrations, and the assignment of a Universal Trial Reference Number to each unique trial worldwide.
- A one-stop search portal for searching registers worldwide.

Thomson CenterWatch (centerwatch.com)

Thomson CenterWatch is a publishing and information services company. Its website, centerwatch.com was among the pioneers in providing a comprehensive list of institutional review board (IRB)-approved clinical trials being conducted internationally. It provides free user access to information about clinical research, including listings of active industry and government-sponsored clinical trials, as well as new drug therapies in research and those recently approved by the FDA.

From attributes to content to benefits:
It's all about the user

Practicing physicians and their patients are likely to be key user groups of clinical trial registers so the value of a register must be measured in three ways: first, by its ability to provide up-to-date information directly related to clinical trial data and relevant to patient treatment; second, by providing knowledge derived from trial data that is easily accessible and understood; and finally, the register's functionality must be intuitive so that anyone experienced with a standard Internet browser can quickly and easily navigate the site.

In addition to practicing physicians and their patients, other potential users span a broad spectrum of individuals and organizations integral to the medical science community. They include (but are not limited to) primary researchers; private, government, charitable and medical research sponsors;

Table 2. Attributes of a good register

- A single, comprehensive Web-based repository accessible through a standard Internet browser
- An assembly of detailed facts and findings of trials from private and public research institutions around the world
- Trial conclusions harnessed for user group applications through the use of parallel clinical trial data articulated and organized using uniform standards
- Clinical relevant analyses that place the results in context
- A layered search function that enables clinical physicians and a wide variety of other user groups, each with differing information needs, to access the type of data, data comparisons, and analyses desired
- Continual updating and maintenance of the Web-based resource

research funders; pharmaceutical companies; drug licensing agencies; medical science publishers; and research ethical review bodies. In addition, access by attorneys representing patients in litigation associated with a specific drug will likely be an audience. To maximize the register's benefit to its users, several attributes should be included (Tab. 2).

Much discussion has taken place regarding the standardized content to be included. The following list represents a comprehensive list based on input from a wide variety of sources. (The least controversial items are listed first.) [4]:

- Title of trial
- Research question
- Study population and interventions
- Lead investigator or institution
- Funding organization
- Unique identifier to prevent repeat registrations
- Other details of methodology – Design, power calculation, outcomes, analysis
- Ethical aspects – Type of consent, information given to participants, approval by ethics committee

- Results – Published or unpublished? Abstracts presented? References to best account of results
- Full protocol

Through the register, trial conclusions and key facts would be publicly accessible. Table 3 highlights the potential benefits to key user groups.

Assessing user needs and the competitive environment

Whether public or privately developed, register developer/providers are experienced disseminators of medical and scientific data products and services. Just as periodic reviews of user needs and the competitive environment are vital to maintaining core products and services, they take on even greater importance at this point in the lifecycle of a relatively new information deliverable as a clinical register. To this end, register developer/providers may wish to consider the following questions:
- Does the clinical register currently or directionally meet its intended purpose?
- Who currently uses the register? When? How often? For what purpose?
- Why should practicing physicians use the register? What is the learning curve for its use?
- What data source(s) are included? How many clinical trials are included in the database? How does the data collected meet the informational, data comparison and analysis needs of clinicians? Is the same type of data collected for all clinical trials?
- How often is the register updated? How is the accuracy of data/findings validated?

Table 3. Potential benefits of clinical trial registers

User group(s)	Content description	Benefit(s)
All	- Purpose of clinical trial and funding source. - Aggregated uniform data about clinical trial findings on a specific subject.	- Satisfy public demand for unbiased evidence on treatment effectiveness. - Promote medical research accountability - Facilitate systematic review of all trial data on a specific subject. - Accelerate dissemination of trial data, making results and potential clinical benefits available sooner. - Discourage publication bias that has the potential to overestimate the efficacy of a particular intervention or give the impression that a certain invention is more promising than the facts might warrant.
Practicing physicians, patients, and patient advocacy groups	- Trial purpose, description, data and findings. - Knowledge derived from aggregated, uniform data.	- Potential for quicker impact on medical practice. - Timely, evidence-based knowledge used for patient treatment.
Researchers, research funders, and ethical review bodies	- Trial purpose, description, data and findings. - Knowledge derived from aggregated, uniform data.	- Informed decision-making. - Avoid unnecessary duplication of effort. - Identify interventions to avoid or reveal potential areas for fruitful investigation. - Save time and money in the drug discovery process. - Foster international collaboration among researchers. - Encourage appropriate replication and confirmation of results. - Identify gaps between research and physician practice. - Enables "research into research."

Table 3. (continued)

User group(s)	Content description	Benefit(s)
		- Provide standardized intelligence on completed trials so funding sources can identify where financial resources are most needed. - Ability to conduct meta-analyses of all trials (regardless of the results) to create systematic reviews of literature on treatments for a particular disease. - Assure close review by relevant ethics committees.

- Is the content of the register duplicative of "competitive" registers? What distinguishes it? What will users remember?
- Do we expect that the register will be a clinician's primary source of data? A back up to journal-published data?
- How will a clinician know about the register and what will drive him/her to it?
- With an increasing number of patients bringing their physicians web-based medical/healthcare information at office visits, is the register's data easily accessible and understood by patients as well as by physicians?
- What is the measure of the register's success in the clinical, regulatory, patient advocacy and medical science environments?

Capturing eyeballs, contending for time

While it is true that the Internet's use by practicing physicians is growing, it cannot be assumed that physicians use it primarily (or secondarily) as a resource to help them find new treatment interventions. Much has been written about the serious time constraints of practicing physicians and their need to find appropriate time balance between patient treatment, practice management and continuing education about medical treatment interventions. Capturing the eyes and holding the attention of clinicians raises several practical issues that bear consideration:

- Of the time a physician will spend learning about medical treatment interventions, what media/resources are considered top priorities?
- What defines a quality, credible and trustworthy media/resource to practicing physicians?
- Are they likely to read peer review/clinical journals during their online experience?
- How does physician age correlate with Internet usage?

Therefore, it is useful also to consider Internet use as well as the medical literature reading habits of practicing physicians. The results of a 2005 survey of office-based physicians on Internet usage revealed [5]: "Almost all physicians have access to the Internet, and most believe it is important for patient care. The most frequent use is in accessing the latest research on specific topics, new information in a disease area, and information related to a specific patient problem. Critical to seeking clinical information is the credibility of the source, followed by relevance, unlimited access, speed, and ease of use. Electronic media are viewed as increasingly important sources for clinical information, with decreased use of journals and local continuing medical education (CME). Barriers to finding needed information include too much information, lack of specific information, and navigation or searching difficulties".

A 2003 Accenture survey of office-based physicians on the biggest influencers of the decision on the drugs they prescribe

Table 4. Results of a study showed respondent ratings of factors that influence their interpretation of medical literature (1 = least to 5 = most important) [7]

Rating	Factor
4.6	Quality of information presented
4.3	How well data support key concepts
4.1	Quality/reputation of journal
3.8	Format (abstract, full article, case study)
3.6	Reputation of organization/institution
3.6	Reputation of investigators
2.7	Commercial sponsorship

found that 80% cited peer review/clinical journals and 16% cited the Internet [6].

A 2004 study of practicing physician-reading habits found that 65% of respondents spent 1–4 h per week reading scientific literature each week. An additional 15% spent 5–6 h per week. The study also revealed respondent ratings of factors that influence their interpretation of medical literature (1 = least to 5 = most important) (Tab. 4) [7].

Driving clinical trial register use and reliance

Despite the good planning and intentions of sponsor/developer organizations, the register environment is fragmented by many factors such as medical specialty, proprietary information and concerns for confidentiality, country, and trial sponsors, etc. The landscape is muddied further by the lack of quality standards for accessing data; content, including accuracy and completeness; inclusion of result summaries/conclusions; language (information edited for academic, legal and/or regulatory purposes not for users); and search criteria and functionality.

In addition, existing registers do not allow for user comparisons between trials or for trial meta-analyses. While much work has been done to bring results of clinical trials to a broad audience, a sizable knowledge gap still remains and represents an enormous opportunity for greater register collaboration.

Understandably, barriers still blur the vision of developing a single-source clinical trial register. To break through, productive collaboration needs to be established between register developers, trial sponsors, regulatory agencies, patient advocacy groups, ethical review bodies, investigator/researchers, and of course patients and their physicians. Confidentiality and intellectual property concerns regarding the impact of public disclosure on the competitive value of proprietary information and trial data must be fully addressed. Uniform standards must be developed for selection/acceptance of clinical trials into the register and the extent of data source/trial content included; accuracy/validation of trial data/conclusions; content logging and information dissemination; formatting of data/search/analytic functions; determining the value of the register to target users; consideration of cultural and language differences within the global environment; and reviewing/synthesizing the best attributes and features of existing registers into the single source. Agreement is needed on where the "Clinical Trial Register Version 2.0" will be maintained and updated; and how will it be funded. This magnitude of effort requires a significant investment of time and resources in the planning, development, and launch stage, and an ongoing commitment to sustain register quality to earn the confidence and trust of users.

The organizing factors for such an endeavor could be summarized in four overarching register attributes:
- Valid, unbiased, uniform content focused on practical benefit to clinicians and patient treatment. Data simply pre-

sented/searchable in the format preferred by physician users. Knowledge acquired through data analysis is easily and strongly related to the data.
- Consistent quality and standards for terminology and for adding/updating trial data. Standardized formats for comparing/analyzing trial data.
- Intuitive user interface easily accessible by anyone capable of using a standard Internet browser.
- Periodic qualitative input by key user groups to assess the practical clinical value and impact of register information and use.

The time is right to take the best of what currently exists and focus its content and functionality on real-world user wants and needs. This effort will result in a robust clinical trial register closer to its original one-source concept, and elevate the sum total of work already accomplished to a higher level of practical clinical relevance and credibility in the medical science community. Ultimately, these factors will drive regular use and reliance.

References

1 DeAngeles CD, Drazen JM, Frizelle FA et al (2004) Clinical Trial registration; a statement from the Internal Committee of Medical Journal Editors. *JAMA* 292: 1363–1364
2 DeAngelis CD, Drazen JM, Frizelle FA et al (2005) Is This Clinical Trial Fully Registered? *JAMA* 293: 2927–2929
3 Simes RH (1986) Publication bias: the case for an international registry of clinical trials. *J Clin Oncol* 4: 1529–1541
4 Tonks A (1999) Registering clinical trials. *BMJ* 319: 1565–1568
5 Bennett NL, Casebeer LL, Kristofco, Strasser SM (2004) Physicians' Internet information-seeking behaviors. *J Conting Ed Health Profess* 24: 31–38

6   Hradecky G (2004) Breaking point: backlash against the number of reps in the field could change the sales model forever. *Pharmaceut Rep*, April 2004

7   Liberti L, Casebeer L et al (2004) Critical appraisal of medical literature by physicians and its relevance to practice. Presented at the 64th Annual Conference of the American Medical Writers Association, St. Louis, MO, October 2004

Appendix

Clinical Trial Registries: A Practical Guide for Sponsors and
Researchers of Medicinal Products, edited by MaryAnn Foote
© 2006 Birkhäuser Verlag Basel/Switzerland

Clinical trial registries and study results databases

MaryAnn Foote

Abraxis BioScience, Inc., 11777 San Vincente Blvd, Suite 550,
Los Angeles, CA 90049, USA

Many clinical trial registries and study results databases were
discussed in the book and several authors discussed the challenges associated with the growing number of Web sites and
databases for these data. Appendix 1 comprises a consolidated
list of registries, databases, and their Web addresses. The lists
are divided into international and government sites (Tab. 1),
oncology cooperative group sites (Tab. 2), and sponsor (academic, commercial, or nonprofit) sites (Tab. 3).

With more than 25 sites available at this time, it is easy to
understand the problems faced by a drug sponsor in terms of
deciding what Web sites to use, if any, besides clinicaltrials.gov,
and the problems faced by a patient in terms of deciding what
Web sites to search about information for a clinical trial for a
particular disease. The problems undoubtedly are increased if
the databases and Web sites are not identical for a given trial.
The plethora of databases will require some form of auditing
to ensure that if an individual trial is listed on multiple registries, the data are uniform and accurate among them. Updating enrollment or site information will need to be done for all
listings of the given trial.

The Web addresses were operative at the time the book
went to press. In the interim, it is possible that some Web sites
are no longer functional, the address has changed, or that other Web sites are available.

Table 1. International and government agency clinical trial and results registries

Name	Web address	Comments
International Portal		
World Health Organization	www.who.int/ictrp	Portal for studies worldwide under development
United States of America		
clinicaltrials.gov	http://www.clinicaltrials.gov	This registry is the largest and most comprehensive clinical trial registry to date. Any research sponsor can register a trial and many sponsors (i.e., trials sponsored by US NIH) are required to register their trials on this site. Results are not posted on this site.
AIDSInfo Clinical Trials	www.aidsinfo.nih.gov	This site is a subset of clinicaltrials.gov. It is sponsored by US NIH specifically for trials that study drugs for HIV and AIDS.
NIH Clinical Trials	http://clinicalstudies.info.nih.gov	This site is a clinical trial registry of studies under the auspices of the US NIH Clinical Center.
Congressionally Directed Medical Research Programs	http://cdmrp.army.mil	The Department of Defense (US) clinical trial registry with links to abstracts and publications based on the clinical trials
Genetic Modification Clinical Research Information System (GeMCRIS)	http://www.gemcris.od.nih.gov	This site is a subset of clinicaltrials.gov. It is sponsored by US NIH specifically for gene therapy trials. Results are given for some studies.

NIH Recombinant DNA Advisory Committee Clinical Trials	www4.od.nih.gov/oba/rac	This site is a subset of clinicaltrials.gov. It is sponsored by US NIH specifically for trials of human gene transfer.
NCI Clinical Trials	www.cancer.gov/clinicaltrials	A clinical trial registry for trials sponsored by the US NCI. A synopsis of clinical trial results also given but not linked to registered trial
NIMH Clinical Trials	www.nimh.nih.gov/studies	A US NIH clinical for the studies sponsored by the NIMH at the NIH
Europe		
Current Controlled Trials	http://www.controlled-trials.com	A UK-based clinical trial registry of studies; an ISRCTN is given for all trials registered here. Results can be published in open-access format and listed on the site.
Dutch Trial Register	http://www.trialregister.nl/trialreg/index.asp	Clinical trial register maintained by Cochrane Center in the Netherlands
ISRCTN site	http://isrctn.org	Clinical trial registry that provides a unique number to randomized controlled trials for all areas of diseases and for all countries worldwide. Administered by Current Controlled Trials Ltd
PENTA Trials	www.ctu.mrc.ac.uk/penta	Clinical trial registry sponsored by the UK Medical Research Council for studies of the treatment of AIDS. Results of studies provided

Table 1. (continued)

Name	Web address	Comments
Rest of World		
Australian Clinical Trial Registry	http://www.actr.org.au	A national register of clinical trials in Australia
National Health and Medical Research Council	www.ctc.usyd.edu.au	Clinical trial registry site maintained by the University of Sydney, Sydney, Australia
University Hospital Medical Information Network	http://www.umin.ac.jp	Clinical trial registry in Japan
South African National Research Register	www.sanrr.gov.za	Clinical trial registry still in development at this time

AIDS = acquired immunodeficiency syndrome; HIV= human immunodeficiency virus; ISRCTN = International Standard Randomised Controlled Trial Number; NCI = National Cancer Institute; NIH = National Institutes of Health; NIMH = National Institute of Mental Health; UK = United Kingdom; US = United States of America

Table 2. Oncology Cooperative Group clinical trial and results registries

Name	Web address	Comments
CALBG	www.calgb.org	A publication registry of CALGB study results. Trials are listed on www.cancer.gov
Gynecologic Cancer Intergroup	http://ctep.cancer.gov	Part of CTEP at NCI, it uses www.clinicaltrials.gov as its clinical trial registry
JCOG	http://www.jcog.jp	Oncology cooperative in Japan
NSABP	www.nsabp.pitt.edu	Clinical trial registry with bibliography of publications based on registered trials
RTOG	www.rtog.org	Clinical trial registry. More than 200 research institutions in the US and Canada below to RTOG, which is affiliated with NCI
SWOG	www.swog.org	Clinical trial registry with bibliography of publications based on registered trials
TrialCheck	www.trialcheck.org	Clinical trial registry sponsored by Coalition of National Cancer Cooperative Groups. Allows searching for phase 3 trials in NCI database

CALGB = Cancer and Leukemia Group B; CTEP = Cancer Therapy Evaluation Program; JCOG = Japan Clinical Oncology Group; NSABP = National Surgical Adjuvant Breast and Bowel Project; RTOG = Radiation Therapy Oncology Group; SWOG = Southwestern Oncology Group

Table 3. Sponsor clinical trial and result registries. Sponsors can be academic, commercial, or nonprofit

Name	Web address	Comments
Amgen	www.amgentrials.com	Clinical trial registry of studies sponsored by Amgen
AstraZeneca	www.astrazeneca-us.com	Clinical trial registry for studies sponsored by AstraZeneca
BRANY	www.brany.com/trials	Clinical trial registry for trials approved by BRANY, a commercial IRB in New York, US
Bristol Myers Squibb	http://ctr.bms.com/ctd	Clinical trial registry for studies sponsored by BristolMyersSquibb
Japan Medical Association	http://dbcentre2.jmacct.med.or.jp/ctrial	Clinical trial registry maintained JMA
CenterWatch	www.centerwatch.com	A commercial clinical trial registry
ClinicalTrials.com	www.clinicaltrials.com	A commercial clinical trial registry
Clinical Trials Search	www.clinicaltrialssearch.org	A commercial clinical trial listing
GlaxoSmithKline	www.gsk.com	Clinical trial and results registry for studies sponsored by GlaxoSmithKline
Hoffman LaRoche	www.roche-trials.com	Clinical trial registry for studies sponsored by Hoffman-LaRoche
IFPMA Clinical Trial Portal	www.ifpma.org/clinicaltrials.html	Clinical trial and results registry for studies. IFPMA is an international organization for drug manufacturers

Japan Pharmaceutical Information Center	http://www.clinicaltrials.jp	Clinical trial registry for studies in Japan
Lilly Trials	www.lillytrials.com	Clinical trial and result registry for studies sponsored by Eli Lilly
Organon	www.organon.com	Clinical trial and result registry for studies sponsored by Organon
PhRMA Clinical Trial Site	www.clinicalstudyresults.org	Clinical trial results registry for marketed products with a package insert for the indication studied. Sponsored by PhRMA

BRANY = Biomedical Research Alliance of New York; IFPMA = International Federation of Pharmaceutical Manufacturers and Associations; IRB = institutional review board; JMA = Japan Medical Association; PhRMA = Pharmaceutical Research and Manufacturers of America

Annotated bibliography of important papers

Tim Peoples and Susan Siefert

Cyberonics, Inc., 100 Cyberonics Blvd, Houston, TX 77058, USA

References were obtained through Entrez PubMed (http://www.pubmed.gov/) using the search terms, "clinical trial [ALL] AND regist* [ALL]". Limits were added to restrict the search to the last 5 years, to eliminate articles not written in English, and to eliminate clinical trial reports from the results. Other references were obtained through manual research.

This bibliography is not, nor is it intended to be, an exhaustive list. News stories, whether from the scientific or popular press, are not included. Relevant letters to the editor published in biomedical journals (marked *LTE* after citation) are included but not summarized. References that could not be obtained are not included.

ARVO statement on registering clinical trials (2006) *Invest Ophthalmol Vis Sci* 47: 1–2
   This policy statement of the Association for Research in Vision and Ophthalmology uses a question and answer format to explain the background and rationale for registering clinical trials.
Expose your clinical thinking (2004) *Nat Biotechnol* 22: 927
   This editorial focuses on the positive aspects of young companies sharing clinical information as a means for obtaining recognition as well as feedback regarding possible flaws in the clinical protocols.
Major pharmaceutical firm concealed information [press release]. New York, NY: Office of the New York State Attorney General Eliot Spitzer; June 2,

2004. Available at: http://www.oag.state.ny.us/press/2004/jun/jun2b_04.html
(Accessed March 17, 2005)

This press release announces a lawsuit by the State of New York against
GlaxoSmithKline (GSK) alleging that the pharmaceutical firm concealed
safety and efficacy information about paroxetine HCl (Paxil), an antidepressant.

Setting new standards (2005) *Nat Rev Drug Discov* 4: 869

This editorial discusses some of the controversies associated with requiring the registration of clinical trials. Among the issues are differences in
the information requested by the Ottawa Group and the World Health
Organization) registries as well as the amount of information that the
pharmaceutical industry is willing to share. Also mentioned is the concept
of pharmaceutical companies depositing proprietary information into a repository for editors to access during peer review with public disclosure of
the results 1 year after the drug has been on the market.

Settlement sets new standard for release of drug information [press release].
New York, NY: Office of New York State Attorney General Eliot Spitzer;
August 26, 2004. Available at: http://www.oag.state.ny.us/press/2004/aug/
aug26a_04.html (Accessed April 19, 2006)

This press release announces a settlement between the State of New York
and GSK. GSK agreed to register all of its clinical trials on its corporate
website and to pay the State of New York $ 2.5 million.

Why should clinical trials be registered? (2005) *CMAJ* 172: 1653, 1655

The editors of *CMAJ* endorse the 2005 ICMJE statement on clinical trial
registration.

Abbasi K (2004) Compulsory registration of clinical trials. *BMJ* 329: 637–638

Abbasi, the editor of the *BMJ*, endorses most of the principles expressed
in the 2005 International Committee of Medical Journal Editors (ICMJE)
statement on clinical trial registration, including mandatory registration of
clinical trials submitted to the *BMJ*. Abbasi, however, suggests that commercial registries, in addition to publicly held registries, should meet the
ICMJE criteria.

Abbasi K, Godlee F (2005) Next steps in trial registration. *BMJ* 330: 1222–
1223

Abbasi and Godlee outline the *BMJ*'s position with regard to the ICMJE's
position that clinical trials be properly registered before the results will be
published in participating journals. At the time this editorial was published,
the *BMJ* had not signed the ICMJE's most recent statement because it
required public ownership and nonprofit status for the trial registry. The
editors maintain that these requirements are too restrictive. As a final note,
they concur with the implication of the ICMJE that 2 or more years may
pass before the success or failure of this effort can be determined.

Antes G (2004) Registering clinical trials is necessary for ethical, scientific and
economic reasons. *Bull World Health Organ* 82: 321
Antes reviews the arguments for and status of clinical trial registration.
The author also expresses support for a WHO decision to register clinical
trials approved by its ethic review board.

Antes G, Dickersin K (2004) Trial registration to prevent duplicate publication. *JAMA* 291: 2432. *LTE*

Bain V (2005) Clinical trials registry. *CMAJ* 172: 979–980. *LTE*

Beppu H (1999) Japan's loss of leadership role in access to drug data. *Lancet*
353: 1992
Beppu reports that a recent change in policy in Japan will allow drug companies to submit unpublished clinical trial data to regulatory authorities,
which was previously disallowed. The author states that Japan should not
have given up its position on access to clinical data; rather, it should have
used its influence to widen access in other countries.

Berlin JA (2005) Why industry should register and disclose results of clinical
studies – perspective of a recovering academic. *BMJ* 330: 959
Berlin discusses the obligation to report results that "indicate harm or lack
of efficacy", and the importance of understanding limitations, especially
those of small and uncontrolled studies.

Besselink MGH, Gooszen HG, Buskens E (2005) Clinical trial registration
and the ICMJE. *JAMA* 293: 157–158. *LTE*

Bonati M, Pandolfini C (2005) More on compulsory registration of clinical
trials: complete clinical trial register is already reality for paediatrics. *BMJ*
330: 480. *LTE*

Bonati M, Pandolfini C (2006) Trial registration, the ICMJE statement, and
paediatric journals. *Arch Dis Child* 91: 93. *LTE*

Bonati M, Pandolfini C, Clavenna A (2004) Disclosure of clinical trials in children. *Science* 305: 1401. *LTE*

Bonati M, Pandolfini C, for the DEC-net Collaborative Group (2005) Pediat-
ric clinical trials registry. *CMAJ* 172: 1159–1160. *LTE*

Braillon A, Dubois G, Slama M (2005) Registration of clinical trials. *Ann Intern Med* 142: 228. *LTE*

Center for Information and Study on Clinical Research Participation. Clinical trial registry survey. Available at: http://www.ciscrp.org/programs/
documents/2005registrysurvey.data.forwebsite2.pdf (Accessed February
8, 2006)
This presentation describes data from a survey investigating the use of
clinical trial registries.

Chalmers I (1990) Underreporting research is scientific misconduct. *JAMA*
263: 1405–1408

Chalmers identifies the underreporting of unfavorable trial results as a form of scientific misconduct. Using examples from perinatal care, the author explains how underreporting can have negative consequences for health systems and patients. The author concludes by calling for expansion of prospective clinical trial registries that include both protocols and results.

Chan A-W, Hróbjartsson A, Haahr MT, Gøtzsche PC, Altman DG (2004) Empirical evidence for selective reporting of outcomes in randomized trials: comparison of protocols to published articles. *JAMA* 291: 2457–2465

Chan et al. conducted a study of the completeness of reporting results of randomized trials approved in Copenhagen and Frederiksberg, Denmark (1994–1995) by comparing study protocols with published articles and surveying trial lists. If the published articles did not contain adequate data for a meta-analysis, the outcome was deemed incompletely reported. The authors found that reports of outcomes frequently provided inadequate data for meta-analysis, showed a bias toward reporting statistically significant findings, and sometimes reported primary outcomes that were inconsistent with those listed in the trial protocols.

Cheng TO (2005) WHO's role of registering trials should include acronyms too. *Int J Cardiol* 102: 369. *LTE*

Couser W, Drüeke T, Halloran P, Kasiske B, Klahr S, Morris P (2005) Trial registry policy: common editorial statement a uniform clinical trial registration policy for journals of kidney diseases, dialysis and transplantation. *Nephrol Dial Transplant* 20: 691[1]

Couser et al., editors of six nephrology and transplantation journals, announce that trials must be registered in a qualified clinical trial registry as a requirement for publication. They provide examples of qualified registries.

Cuellar D (2005) Several clinical-registry issues should concern practicing pharmacists. *J Am Pharm Assoc (Wash DC)* 45: 10

Cuellar reviews issues arising from clinical trial registration that are relevant to practicing pharmacists.

Curfman GD, Morrissey S, Drazen JM (2005) Expression of concern: Bombardier et al., "Comparison of upper gastrointestinal toxicity of rofecoxib and naproxen in patients with rheumatoid arthritis," *N Engl J Med* 2000;343: 1520–1528. *N Engl J Med* 353: 2813–2814

The editors of the *New England Journal of Medicine* express concern after a memorandum suggests data were concealed in a report about the pain-

---

1   Similar editorials appeared in other journals (*Transplantation*. 2005;79:751; *J Am Soc Nephrol* 2005;16:837; and *Am J Transplant* 2005;5:837). The editors of the *American Journal of Kidney Diseases and Kidney International* signed the editorial, but it did not appear in either journal.

killer rofecoxib (Vioxx). The editors ask that the authors submit a correction.

D'Alonzo GE Jr (2004) Clinical trial registration, a needed addition to the research process. *J Am Osteopath Assoc* 104: 409–410

D'Alonzo, editor of the *Journal of the American Osteopathic Association*, endorses the 2004 ICMJE statement on clinical trial registration.

De Angelis C, Drazen JM, Frizelle FA et al. Clinical trial registration: a statement from the International Committee of Medical Journal Editors. Available at: http://www.icmje.org/clin_trial.pdf (Accessed January 29, 2006)[2]

The ICMJE members state that all clinical trials published in their journals must be registered before publication. The ICMJE members encourage other biomedical journals to publish and follow these guidelines.

De Angelis C, Drazen JM, Frizelle FA et al. Is this clinical trial fully registered? A statement from the International Committee of Medical Journal Editors. Available at: http://www.icmje.org/clin_trialup.htm (Accessed January 29, 2006)[2]

The ICMJE members update their previous statement in response to questions from journal editors and in light of WHO meetings. All clinical trials with at least one prospectively assigned control or comparison group must be registered. Also, all registered clinical trials must fulfill the minimum 20 data sets devised by the WHO.

Decullier E, Lhéritier V, Chapuis F (2005) Fate of biomedical research protocols and publication bias in France: retrospective cohort study. *BMJ* 331: 19–22

Decullier et al. report data from a national survey examining publication bias in France. The survey found that confirmatory results predicted publication, confirmatory results significantly led to less time to publication, and a low percentage of completed studies were published. The authors conclude that their data provide evidence of publication bias and that prospective clinical trial registration is necessary. They state that this survey is the first national study of publication bias.

Della Cioppa G, Garaud J-J (2003) Registering clinical trials. *JAMA* 290: 2545–2546. *LTE*

Dickersin K, Davis BR, Dixon DO et al. (2004) The Society for Clinical Trials supports United States legislation mandating trials registration. *Clin Trials* 1: 417–420

---

2 Published simultaneously in all signatory journals (*JAMA, New England Journal of Medicine, The New Zealand Medical Journal, Norwegian Medical Journal, CMAJ, Lancet, Annals of Internal Medicine, Croatian Medical Journal, Nederlands Tijdschrift voor Geneeskunde, Journal of the Danish Medical Association,* and *The Medical Journal of Australia*). Subsequently published in many other biomedical journals.

The Society for Clinical Trials (SCT), represented by Dickersin et al., endorses national legislation and funding for a centralized clinical trial registry. The SCT proposes a standard for registration and a minimum data set, both of which are similar to those proposed by the ICMJE. Furthermore, the SCT endorses legislation requiring clinical trial registration.

Dickersin K, Rennie D (2003) Registering clinical trials. *JAMA* 290: 516–523
Dickersin and Rennie review the status and history of clinical trial registration, explain barriers to adoption of registration, and offer recommendations to further registration. The authors identify barriers from industry, lawmakers, and the press. To advance registration, the authors recommend NIH leadership, enforcement by institutional review boards, active support by industry, the use of unique registration numbers, and pressure from lawmakers.

Drazen JM, Wood AJJ (2005) Trial registration report card. *N Engl J Med* 353: 2809–2811
Drazen and Wood comment on a report by Zarin et al. about ClinicalTrials.gov. The authors focus on industry noncompliance with legislation requiring clinical trial registration and ICMJE statements.

Drennan KB (2005) Registration of clinical trials. *Contemp Clin Trials* 26: 517
Drennan speculates about problems that may result from providing full access to trial data, many of which are related to how the public will perceive and interpret the technical material. The editorial concludes with an affirmation that the journal will endeavor to publish "the best quality scholarly articles."

Dresser R (2005) A new era in drug regulation? *Hastings Cent Rep* 35(3): 10–11
Dresser reviews and advocates arguments for mandatory clinical trial registration and expanded post-marketing surveillance of medical products.

Dresser R (2005) Clinical trial registration and the ICMJE. *JAMA* 293: 157–158. *LTE*

Easterbrook PJ, Berlin JA, Gopalan R, Matthews DR (1991) Publication bias in clinical research. *Lancet* 337: 867–872
Easterbrook et al. examine publication bias in all studies approved by a British research ethics board over 4 years. The authors report that statistically significant results predicted both publication and acceptance in high-profile journals. The authors propose clinical trial registration to curb publication bias.

Evans T, Gülmezoglu M, Pang T (2004) Registering clinical trials: an essential role for WHO. *Lancet* 363: 1413–1414
Evans et al. state the importance of clinical trial registration to healthcare in middle- and low-income countries. The authors call for an international

registry or meta-registry. They also applaud recent and future WHO efforts to expand clinical trial registration.

Eysenbach G (2004) Tackling publication bias and selective reporting in health informatics research: register your eHealth trials in the International eHealth Studies Registry. *J Med Internet Res* 6: e35
Eysenbach introduces the International eHealth Studies Registry as an alternative to ClinicalTrials.gov. The author notes that the new registry will be ideal for eHealth studies and for studies conducted outside the United States.

Fair Access to Clinical Trials Act of 2004, US House of Representatives, HR 5252, 108th Congress, 2nd Session. Available at: http://www.govtrack.us/data/us/bills.text/108/h5252.pdf (Accessed February 9, 2006)
This bill would require all sponsors, whether private or public, of clinical trials of drugs, biologics, or medical devices to register both study details and results in a federally mandated database. The bill also gives enforcement power to the Secretary of Health and Human Services.

Fair Access to Clinical Trials Act of 2004, US Senate, S 2933, 108th Congress, 2nd Session. Available at: http://www.govtrack.us/data/us/bills.text/108/s2933.pdf (Accessed February 9, 2006)
This bill would require all sponsors, whether private or public, of clinical trials of drugs, biologics, or medical devices to register both study details and results in a federally mandated database. The bill also gives enforcement power to the Secretary of Health and Human Services.

Falit B (2005) Pharma's commitment to maintaining a clinical trial register: increased transparency or contrived public appeasement? *J Law Med Ethics* 33: 391–396
Falit analyzes the joint registry proposed by four major pharmaceutical associations. The author concludes that industry commitment to clinical trial registration is inconsistent and further legislation is needed to ensure compliance.

Feczko J (2006) Clinical trials report card. *N Engl J Med* 354: 1426–1429. *LTE*

Fisher CB (2006) Clinical trial results databases: Unanswered questions. *Science* 311: 180–181
Fisher discusses possible effects of clinical trial registration on scientific peer review, the patient-doctor relationship, insurance coverage, and industry competition. The author states that while prompt dissemination of results is necessary, caution must be exercised in registering clinical trials.

Foote MA (2004) Manuscripts based on clinical trial results: dilemmas of medical writers. *AMWA J* 19: 158
Foote reports an open session in which bias and authorship issues were discussed.

Foote MA (2003) Review of current authorship guidelines and the controversy regarding publication of clinical trial data. *Biotechnol Annu Rev* 9:
303–313
  Foote summarizes the controversies surrounding and policies applied to
  authorship of multicenter studies by multiple authors, including academic
  institutions and drug sponsors.

Foote MA, Noguchi PD (2005) Posting of clinical trials and posting of clinical
trial results: information for medical writers. *AMWA J* 20: 47-49
  Foote and Noguchi review recent actions and policies regarding clinical
  trial registration enacted by regulatory agencies, journal editors, and industry associations. The authors also review the status of clinical trial registries. The authors state that clinical trial registration will improve industry
  efficiency and patient access.

Gøtzsche PC (2005) Research integrity and pharmaceutical industry sponsorship. Trial registration, transparency and less reliance on industry trials are
essential. *Med J Aust* 182: 549–550
  Gøtzsche discusses industry influence over clinical trial data and publication bias. The author calls for procedures that would curb industry influence and clinical trial registration.

Grass G (2005) Clinical trial registration. *N Engl J Med* 352: 198–199. *LTE*

Gülmezoglu AM, Pang T, Horton R, Dickersin K (2005) WHO facilitates international standards for clinical trial registration. *Lancet* 365: 1829–1831
  Gülmezoglu et al. review discussions at a WHO meeting about clinical trial
  registration held in April 2005.

Gülmezoglu AM, Sim I (2006) International clinical trial registration: Any
progress? *Wien Klin Wochenschr* 118: 1–2
  Gülmezoglu and Sim provide an update on WHO discussions about clinical trial registration.

Habibzadeh F (2006) Call for mandatory registration of clinical trials and its
impact on small medical journals: scenario on emerging bias. *Croat Med J*
47: 181–183. *LTE*

Haug C, Gøtzsche PC, Schroeder TV (2005) Registries and registration of
clinical trials. *N Engl J Med* 353: 2811–2812
  Haug et al. discuss the current state of clinical trial registration and advocate removal of barriers to wider registration.

Hirsch LJ (2002) Conflicts of interest in drug development: the practices of
Merck & Co. Inc. *Sci Eng Ethics* 8: 429–442
  Hirsch reviews drug development practices of Merck. The author states
  that Merck does not disclose the design of all of its clinical trials at inception because the company needs to protect proprietary information.

Hoffer LJ, Mendelson J (2005) Clinical trials registry. *CMAJ* 172: 980. *LTE*

Horton R, Smith R (1999) Time to register randomised trials. The case is now unanswerable. *BMJ* 319: 865–856

Horton and Smith state that progress has been made in clinical trial registration, but much more needs to be accomplished.

Jay P, Wallace M (2004) Compulsory registration of clinical trials: under-reporting is not an option. *BMJ* 329: 1044. *LTE*

Johnson K, Lassere M (2006) Clinical trials report card. *N Engl J Med* 354: 1427-1428. *LTE*

Jull A, Wills M, Scoggins B, Rodgers A (2005) Clinical trials in New Zealand – treading water in the knowledge wave? *N Z Med J* 118: U1638

Jull et al. used records in the public domain to collect the same types of information about clinical trials that would be contained in a clinical trials registry. They note that in New Zealand, the ethics committees that approve clinical trials would serve as an excellent resource to help develop a clinical trials registry.

Kennedy D (2004) The old file-drawer problem. *Science* 305: 451

After delineating how positive results are more likely to be published than negative results, Kennedy praises the efforts by journal editors to encourage registration of clinical trials and cites current events that may have influenced the trend.

Khalil O, Govindarajan R, Safar M, Hutchins L, Mehta P (2005) Clinical trial registration and the ICMJE. *JAMA* 293: 157–158. *LTE*

Krall R, Rockhold F (2005) More on compulsory registration of clinical trials: GSK has created useful register. *BMJ* 330: 479–480. *LTE*

Krall R, Rockhold F (2005) Trial registration: ignored to irresistible. *JAMA* 293: 158. *LTE*

Krall RL, Rockhold F (2006) Clinical trials report card. *N Engl J Med* 354: 1427–1429. *LTE*

Krleža-Jerić K (2005) Clinical trial registration: the differing views of industry, the WHO, and the Ottawa group. *PLoS Med* 2: e387

Krleža-Jerić reviews the history and status of clinical trial registration after an April 2005 WHO meeting. The author describes the approaches to clinical trial registration taken by industry, the WHO, and the Ottawa Group.

Krleža-Jerić K, Chan A-W, Dickersin K, Sim I, Grimshaw J, Gluud C, for the Ottawa Group (2005) Principles for international registration of protocol information and results from human trials of health related interventions: Ottawa statement (part 1). *BMJ* 330: 956–958

Krleža-Jerić et al. outline the ethical and scientific rationale as well as principles of part 1 of the Ottawa statement on registration of clinical trials. Emphasizing the social contract "to accurately disseminate information from all trials," the authors address several issues. With regard to ineffective or harmful interventions, they note that making such information available can help prevent duplication. To counter pharmaceutical industry concerns about revealing trade secrets, they point out sources that already disseminate such information and note that design properties are not required for registration. They also caution readers to distinguish between peer-reviewed and non-peer-reviewed information, especially considering the possible availability of unpublished data.

Kulvichit K, Kulwichit W, Lumbiganon P (2005) Clinical trial registration. *N Engl J Med* 352: 198–199. *LTE*

LeBlanc JC (2005) Clinical trials registry. *CMAJ* 172: 980. *LTE*

Leviton I (2003) Registering clinical trials. *JAMA* 290: 2545–2546. *LTE*

Levin LA, Gottlieb JL, Beck RW et al (2005) Registration of clinical trials. *Arch Ophthalmol* 123: 1263–1264

Levin et al. explain the requirements for clinical trial registration for *Archives of Ophthalmology*, *American Journal of Ophthalmology*, *British Journal of Ophthalmology*, and *Ophthalmology*. They discuss their policy of refusing to review trials that were not registered unless the results of the unregistered trials are of great importance to the public health. They justify the requirement for clinical trial registration by citing the risks assumed by the subjects and the ethical injustice of failure to publish negative results and adverse effects.

Levy JA, Autran B, Coutinho R, Phair JP (2005) Registration of clinical trials. *AIDS* 19: 105

Levy et al. announce that clinical trials published in *AIDS* must be registered in a clinical trial registry. They emphasize the goal of the ICMJE: complete, rather than selective reports of clinical trials. They justify the requirement for full reporting by citing the commitment and contribution of the volunteer participants.

Lexchin J (2004) Clinical trials register. *Lancet* 364: 330. *LTE*

Lie F (2004) Registering clinical trials. *Lancet* 363: 2191. *LTE*

MacDonald D, Molnari PM (2005) Breaking the trial result disclosure logjam now. *Appl Clin Trials* 11: 37–38

MacDonald and Molnari state that current industry efforts to register clinical trials are making progress but are being pushed aside by lawmakers. The authors call for cooperation between industry and lawmakers to expand clinical trial registration.

Maine Department of Health and Human Services. Clinical Trial Registration Law Summary. Available at:
http://www.maine.gov/dhhs/boh/clinical_trials.htm (Accessed February 16, 2006)

This website summarizes a recent Maine law that requires clinical trial registration of pharmaceutical companies advertising in Maine.

Manheimer E, Anderson D (2002) Survey of public information about ongoing clinical trials funded by industry: evaluation of completeness and accessibility. *BMJ* 325: 528–531
    Manheimer and Anderson searched for prostate and colon cancer drugs in industry pipeline sources and clinical trial registries. The authors found both sources inconsistent and incomplete.

Manrow RE (2005) Re: Clinical trials registration efforts gain some ground. *J Natl Cancer Inst* 97: 936. *LTE*

McCormack J, Loewen P, Jewesson P (2005) Dissemination of results needs to be tracked as well as the funding is. *BMJ* 331: 456. *LTE*

McCray AT (2000) Better access to information about clinical trials. *Ann Intern Med* 133: 609-614
    McCray reviews the status of clinical trial registration and describes ClinicalTrials.gov.

Moher D, Bernstein A (2004) Registering CIHR-funded randomized controlled trials: a global public good. *CMAJ* 171: 750–751
    Moher and Bernstein endorse a decision by the Canadian Institutes of Health Research to register all the clinical trials that it funds.

Ormerod AD, Williams HC (2005) Compulsory registration of clinical trials. *Br J Dermatol* 152: 859–860
    Emphasizing the value of access to all trials for the purpose of conducting reviews, Ormerod and Williams point out 2 additional advantages of clinical trial registration: discouraging post hoc subgroup analyses and duplication of research efforts.

Pengel L, Barcena L, Morris PJ (2005) Registry of randomized controlled trials in transplantation. *Transplantation* 80: 432
    Pengel et al. explain the criteria for selection and rating of published randomized controlled trials on solid organ transplantation featured in a registry published semi-annually in *Transplantation*. This registry concentrates on the methods used in the trial and lists a brief aim and rating of the trials. The Jadad scale is used to assess the quality of the trial. Concealment of allocation and intent to treat analyses comprised two additional aspects of quality.

Pharmaceutical Research and Manufacturers of America. Joint position on the disclosure of clinical trial information via clinical trial registries and databases. Available at: http://international.phrma.org/publications/policy//admin/2005-01-06.1113.PDF (Accessed February 8, 2006)
    This joint statement of 4 international pharmaceutical industry associations commits to registering clinical trials. The statement also affirms the industry's right to protect proprietary information.

Pharmaceutical Research and Manufacturers of America. PhRMA Clinical Study Results Database Proposal. Available at:

http://www.clinicalstudyresults.org/primers/Clinical_Study_Results_ Database.pdf (Accessed February 8, 2006)
  This document proposes the framework for ClinicalStudyResults.org, PhrMA's clinical trial registry.
Reidenberg MM (2006) Clinical trials report card. *N Engl J Med* 354: 1428– 1429. *LTE*
Rennie D (2004) Trial registration: a great idea switches from ignored to ir- resistible. *JAMA* 292: 1359-1362
  Rennie reviews industry reluctance to register clinical trials. The author calls for a national registry of all clinical trials, whether privately or pub- licly sponsored, that is not connected to industry.
Reveiz L, Cardona AF, Ospina EG (2005) Clinical trial registration. *N Engl J Med* 352: 198–199. *LTE*
Roberts I (1998) An amnesty for unpublished trials. One year on, many trials are unregistered and the amnesty remains open. *BMJ* 317: 763-764
  Roberts reviews the response to an amnesty on unpublished clinical trials announced by 50–100 biomedical journals. Few trials were registered after 1 year of amnesty, but the author notes that the journals' stance had a marked effect on dialogue about clinical trial registration.
Sammons HM, Naylor C, Choonara I, Pandolfini C, Bonati M (2005) Should paediatricians support the European Paediatric Clinical Trials Register? *Arch Dis Child* 90: 559–560
  Sammons et al. present the rationale and design of the European Paediat- ric Clinical Trials Register.
Sim I, Detmer, DE (2005) Beyond trial registration: a global trial bank for clinical trial reporting. *PLoS Med* 2: e365
  Sim and Detmer introduce the Global Trial Bank, a clinical trial registry developed by the American Medical Informatics Association and connect- ed with a new online journal, *PLoS Clinical Trials*.
Simes RJ (1986) Publication bias: the case for an international registry of clinical trials. *J Clin Oncol* 4: 1529–1541
  Simes analyzes trials of oncology treatments and finds notable differences between published and registered trials. The author suggests an interna- tional clinical trial registry as a tool to overcome the gaps in ordinary lit- erature searches. Simes is frequently credited as the first to suggest this concept.
Singer EA, Druml C (2004) Compulsory registration of clinical trials: maybe European research should be protected. *BMJ* 329: 1044. *LTE*
Somberg J (2004) Drug trial registries. *Am J Ther* 11: 327
  Somberg points out major flaws in the move to require that clinical tri- als be registered before their results will be considered for publication. An ironic observation is that clinical trial registries will allow competing

companies to keep track of each other's activities. The author reminds the reader that this particular journal has historically acknowledged the importance of negative clinical studies. Somberg notes that the registries will not serve a purpose in making therapeutic decisions, but may inappropriately serve to impose an embargo on publishing unregistered trials.

Staessen JA, Bianchi G (2003) Registration of trials and protocols. *Lancet* 362: 1009–1010

Staessen and Bianchi examine how clinical trial registration can reduce redundant and ill-conceived research. The authors identify registration of trial protocols at inception as an essential element in clinical trial registration.

Steinbrook R (2004) Public registration of clinical trials. *N Engl J Med* 351: 315–317

Steinbrook reviews the arguments for and status of clinical trial registration.

Steinbrook R (2004) Registration of clinical trials – voluntary or mandatory? *N Engl J Med* 351: 1820–1822

Steinbrook reviews attempts to advance and impede mandatory clinical trial registration.

Stern JM, Simes RJ (1997) Publication bias: evidence of delayed publication in a cohort study of clinical research projects. *BMJ* 315: 640–645

Stern and Simes examined publication bias and delay in studies submitted to an Australian ethics committee. The authors found that a statistically significant outcome predicted publication. Publication delay was associated with lack of a statistically significant outcome. The authors advocate prospective clinical trial registration as a solution to these problems.

Stewart L, Vale C, Darbyshire J (2004) Compulsory registration of clinical trials: publicly funded national register of trials would be best in the United Kingdom. *BMJ* 329: 1043–1044. *LTE*

Stone JH, Lockshin MD, Katz PP, Yelin EH (2005) New journal policy regarding registration of clinical trials. *Arthritis Rheum* 52: 2243–2347

Stone et al. review the arguments for and against clinical trial registration and announce adoption of the 2004 and 2005 ICMJE statements in both journals of the American College of Rheumatology, *Arthritis & Rheumatism* and *Arthritis Care & Research*.

Sundar S, Lawton P (2004) International register of trial acronyms. *Lancet* 363: 171. *LTE*

Tamir O, Lipschitz Y, Shemer J (2006) Clinical trials report card. *N Engl J Med* 354: 1427–1428. *LTE*

Tonks A (1999) Registering clinical trials. *BMJ* 319: 1565–1568

Tonks reviews the status and future directions of clinical trial registration. The author bases the article on discussions at a conference sponsored by the *BMJ*, *The Lancet*, and the Association of the British Pharmaceutical Industry.

Tonks A (2002) A clinical trials register for Europe. *BMJ* 325: 1314–1315
Tonks reviews challenges facing clinical trial registration following discussions at a November 2002 meeting of the European Science Foundation.

United States Food and Drug Administration. Food and Drug Administration Modernization Act of 1997, Section 113. Available at: http://lhncbc.nlm.nih.gov/clin/113.html (Accessed February 8, 2006)
This act directs that a clinical trials registry (ClinicalTrials.gov) be initiated and maintained for all studies of experimental treatments for serious or life-threatening conditions.

Van Der Weyden MB, Ghersi D (2005) The Australian Clinical Trial Registry. *Med J Aust* 183: 7
Van Der Weyden and Ghersi explain attempts by the ICMJE to advance clinical trial registration and present the Australian Clinical Trial Registry.

World Health Organization. WHO technical consultation on clinical trial registration standards. April 25–27, 2005; Geneva, Switzerland. Available at: http://www.who.int/ictrp/news/ictrp_sag_meeting_april2005_conclusions.pdf (Accessed January 30, 2006)
This document explains conclusions reached at a WHO meeting on clinical trial registration. This meeting led to the 20-item minimum data set eventually endorsed by the ICMJE.

Zarin DA (2006) Clarifying a misunderstanding on clinical trial registry. *CMAJ* 174: 203, 206. *LTE*

Zarin DA (2005) Clinical trial registration. *N Engl J Med* 352: 1611. *LTE*

Zarin DA, Tse T, Ide NC (2005) Trial registration at ClinicalTrials.gov between May and October 2005. *N Engl J Med* 353: 2779-2787
Zarin et al. provide an update on registration at ClinicalTrials.gov. The authors note a marked increase in registration shortly before and shortly after the ICMJE policy on clinical trial registration went into effect. While the authors note signs of progress, they express concern at the tendency toward incomplete responses.

Index